PROFESSIONAL BREAD BAKING

HANS WELKER
The Culinary Institute of America
Erin Jeanne McDowell

PROFESSIONAL BREAD BAKING

Photography by Jennifer May

WILEY

This book is printed on acid-free paper. ∞

THE CULINARY INSTITUTE OF AMERICA

PRESIDENT	Dr. Tim Ryan '77, CMC
PROVOST	Mark Erickson '77, CMC
DIRECTOR OF PUBLISHING	Nathalie Fischer
SENIOR EDITORIAL PROJECT MANAGER	Lisa Lahey '00
RECIPE TESTING MANAGER	Laura Monroe '12

Photography by Jennifer May and Phil Mansfield

Published by John Wiley & Sons, Inc., Hoboken, New Jersey

Published simultaneously in Canada

For general information on our other products and services, or technical support, please contact our Customer Care Department within the United States at 800-762-2974, outside the United States at 317-572-3993 or fax 317-572-4002.

Wiley publishes in a variety of print and electronic formats. Some content that appears in print may not be available in electronic books. For more information about Wiley products, visit our website at www.wiley.com.

Cover Design by Thomas Nery
Interior Design by Wendy Lai

Library of Congress Cataloging-in-Publication Data:

Names: Welker, Hans, author.
Title: Professional bread baking / Hans Welker.
Description: Hoboken, New Jersey : John Wiley & Sons, Inc., [2016] | Includes bibliographical references and index.
Identifiers: LCCN 2015043238 (print) | LCCN 2015046486 (ebook) | ISBN 9781118435878 (cloth : acid-free paper) | ISBN 9781118435847 (pdf) | ISBN 9781118435854 (epub)
Subjects: LCSH: Bread.
Classification: LCC TX769 .W387 2016 (print) | LCC TX769 (ebook) | DDC 641.81/5—dc23
LC record available at http://lccn.loc.gov/2015043238

Printed in the United States of America

SKY10076881_060624

CONTENTS

Recipe Contents

7 Flatbreads and Breads from Around the World

8 Rye Breads and Rolls

Preface

UNDERSTANDING BREADS

Though the popularity of bread has rarely waivered worldwide over the centuries, an appreciation of true artisan products has fluctuated through history. While the production of high-quality loaves was on a steady rise well into the twentieth century, the advent of industrialization took a toll on the prevalence of artisan bread techniques. Fortunately, bakers throughout the world have returned to these prized techniques, and through their efforts have put renewed emphasis on the importance of handmade products using ages-old traditions.

Professional Bread Baking is a compendium of information on this subject, fully covering the process of bread baking from its early development in ancient Egypt to the techniques of today's artisans. Chapter 1 gives a detailed history of bread from its initial creation to a survey of the varieties available today. These historical accounts also cover in greater detail those countries that particularly place an emphasis on bread production, including France, Italy, Germany, and eastern Europe. Additionally, this chapter discusses the role of bread in religion, folklore, and other traditions throughout the world.

Chapter 2 explains the equipment used to produce bread, while Chapter 3 offers descriptions of the key ingredients used in the baking process. Chapter 4 provides expansive details on the bread-baking process from start to finish.

The remaining chapters in *Professional Bread Baking* include a variety of recipes for various types of breads. For instance, Chapter 5 provides recipes for the production of lean dough breads and rolls. Chapter 6 contains recipes for breads and rolls made with enriched dough. Chapter 7 is full of recipes for flatbreads and other unique breads from around the world. And Chapter 8 delves into the process of making artisan breads from one of the most complex grains and flours: rye.

A History of Bread

The importance of bread through the centuries cannot be overstated. From the first rustic types produced almost 6,000 years ago, bread has been at the center of incredible moments in history. As one of the most prominent food sources in the world, it has become a symbol for prosperity, health, wealth, and success. It found its way into religious ceremonies, fairy tales, folklore, and regional traditions. The production of bread encouraged technological advances in agriculture, as inventors worked to create new tools to harvest grain more efficiently. Bread even played an important role in the natural sciences, as soil was studied and grains were cross-bred to produce higher yields and heartier varieties. Additionally, bread has served to divide the classes, while its vital importance to all people led to revolution in times of famine. This history provides new insight into how and why bread is produced the way it is today.

The Right Tools for the Job

The tools needed to produce artisan bread are relatively minimal. Many bakers will speak at great length about the importance of their hands and their own experience as the most important factors for making excellent loaves. However, when it comes to tools, choosing the right ones is incredibly important. This book goes into detail about every type of tool commonly used for producing artisan bread in a professional bakeshop. Beyond this, some pros and cons are listed for large items—items like mixers or ovens that can be major purchases for a bakeshop. Even if not purchasing new equipment, bakers can better understand *how* to use the equipment they currently have, as well as make compensations to produce the best product possible with existing equipment. Chapter 2 delves into this subject, providing new insights into the production of the recipes that follow.

Detailed Methods and Techniques

Bread dough is a living thing—it breathes and is changing constantly from the moment the ingredients are mixed until the loaves are removed from the oven. From a relatively simple and minimal set of ingredients, a huge number of varied products can be formed. It is vital, thus, to understand the steps and techniques required to make bread so as to end up with the desired result.

Chapter 3 discusses ingredients, providing insights into how they should be treated, and describes how they work together to form the finished loaf. Chapter 4 covers the twelve steps of bread baking, describing each step of the process. These details are critical for understanding bread production, and they provide an excellent foundation for acquiring the skill set needed to be an artisan baker—a skill set that develops and is perfected by time and experience. Most important, the chapter offers a deep understanding of the methods that produce consistently great bread, which is vital to attracting customers and maintaining repeat business.

The Importance of Regionality

One of the things that sets this book apart from others on the subject is the information included on the regional shapes characteristic of prominent bread-baking countries like France, Italy, Germany, and others. Throughout history, bread has been used to commemorate kingdoms, towns, or villages. The people of these localities took great pride in a product that made their regions unique. Displaying a range of ingredients, shapes, methods, and techniques, regional loaves reflect a rich bread tradition that has often gone unnoticed. This has been remedied in this volume, with its discussion of the historical significance of some of these regional products, to the formulas that reproduce these artisan bread and rolls.

The Greatness of Rye

Rye is one of the most misunderstood grains, especially when it is ground into flour and used in bread baking. Many people don't realize that some of the most popular loaves have used rye flour, often mixed with wheat flour to produce a deeply flavorful loaf. Rye grains are structured differently from wheat grains, producing a flour with less ability to form a strong gluten structure. If rye flour is used improperly, it can yield an overly dense loaf. But understood properly, with techniques such as incorporation of a sour starter, rye flour can produce truly amazing artisanal products. Thus, in addition to providing plenty of information on rye flour and how to use it, Chapter 8 has recipes featuring rye.

Book Features

This book is not only a cookbook providing an array of recipes and formulas for finished loaves. It also features discussions about bread, becoming a detailed reference that will be indispensible for any baker. Whether you are looking to integrate a bread program into a restaurant, hotel, or bakery, or your aim is to streamline production at a current bread-specific bakeshop, this volume offers the tools you need to mix, ferment, shape, proof, and bake exceptional artisanal bread.

Resources

Professional Bread Baking offers an **Instructor's Manual,** including a **Test Bank,** to help instructors who are designing courses based on healthy menu items.

A password-protected Wiley Instructor Book Companion Website (www.wiley.com/college/CIA) provides access to the online **Instructor's Manual** and the text-specific teaching resources. The **Power Point slides** are also available on the Website for download.

Acknowledgments

Writing a book is easy. That is, until you start the process. This book would never have happened without the help of so many people. I would like to thank the CIA family, who so powerfully influence my life, my work, and this book.

A special thanks goes to the publishing department at the CIA: Maggie Wheeler, whose encouragement and direction kept me on track; Lisa Lahey, for her help in making this project a learning tool; Laura Monroe's spirit lifted me up so many times—she went from a former student to my "book instructor"; and Erin McDowell, for her tireless pursuit of accuracy and clarity in my notes.

Thanks to Kevin Curtin, Sun ah Kim, Vivian Jago, Marit Rubenstein, and Nancy Swift for their help in putting my recipes into the correct format. Thank you to my weekend photo shoot assistants, Monica Hannoush and Kelsey Williamson.

This book would be nothing without the artful photography taken by Jennifer May and her assistant, Kazio Sosnowski, during the countless weekends we spent in the Breads bakeshop. She will be a better baker and I will take pictures with more "focus."

My warmest thanks to Phil Mansfield; his creative pictures tremendously improved this book.

And finally, I would like to thank my sons: Carl for his inspiring feedback; Greg and Lars; my father-in-law, Jody; and the love of my life, my wife and best friend, Marie Elaina. I am nothing without you. Lastly, this book would be nowhere if not for my two bulldogs, Pimp and Lola, who sat next to me for hours as I was fighting with the computer.

About the Authors

HANS WELKER

Hans Welker, CMB, CHE, is an associate professor of baking and pastry arts at The Culinary Institute of America (CIA). Chef Welker currently teaches Specialty Breads for baking and pastry arts majors in the college's degree programs. Chef Welker's students learn the principles and techniques of preparing multigrain breads, sourdoughs, bagels, pretzels, holiday and seasonal breads, and flatbreads. His class emphasizes regional breads and breads of the world, as well as innovative baking methods.

Before joining the CIA faculty in 2006, Chef Welker was the director of the bread program at the French Culinary Institute in New York City. He was also the owner, baker, and master pastry chef at Pastry Paradise Konditorei and the Alpine Bakery & Café in Lagrangeville, New York; head pastry chef at Karl Ehmer's Bakery in Poughkeepsie, New York, and at Rathauscafe Harth and Konditorei Späth in Darmstadt, Germany; and pastry chef in Münster and Idar-Oberstein, Germany.

Hans Welker is a Certified Master Baker (CMB) and Certified Hospitality Educator (CHE). He completed his apprenticeship at Café Benner in Idar-Oberstein and served as a pastry chef in the German Navy. He holds a master's degree from the Independent University of Heidelberg's Gemeinnützige Fortbildung Schule and, since 1988, has been a German Master Pastry Chef. He also studied at the Ewald Notter International Zuckerdecorschule in Zurich, Switzerland; Moll Marzipan in Berlin, Germany; and The Coalition of Teachers in New York City.

Chef Welker was a member of the CIA faculty team that in 2007 won the Marc Sarrazin Cup Grand Prize from the Société Culinaire Philanthropiqué at the New York Salon of Culinary Art. At both the 2009 and 2010 New York Salons, he earned the Joseph Donon Medal of the Auguste Escoffier Foundation for Best Bakery Display, and in both 2011 and 2012, he won a gold medal and Best of Show award for his bread display.

THE CULINARY INSTITUTE OF AMERICA

Founded in 1946, The Culinary Institute of America is an independent, not-for-profit college offering associate and bachelor's degrees with majors in culinary arts, baking and pastry arts, and culinary science, as well as certificate programs in culinary arts and wine and beverage studies. As the world's premier culinary college, the CIA provides thought leadership in the areas of health and wellness, sustainability, and world cuisines and cultures through research and conferences. The CIA has a network of 45,000 alumni that includes industry leaders such as Grant Achatz, Anthony Bourdain, Roy Choi, Cat Cora, Dan Coudreaut, Steve Ells, Johnny Iuzzini, Charlie Palmer, and Roy Yamaguchi. The CIA also offers courses for professionals and enthusiasts, as well as consulting services in support of innovation for the food service and hospitality industry. The college has campuses in Hyde Park, New York; St. Helena, California; San Antonio, Texas; and Singapore.

AN INTRODUCTION TO ARTISAN BREADS

1

Throughout the world, bread is regarded as an essential food. It is the cornerstone of almost every diet. It has played a role in the histories and cultures of dozens of countries. It has a place in folklore, traditions, and religions. But beyond that, how bread is made, sold, and eaten is truly indicative of the place it has come from and of the people who live there. It has been an incredibly important part of people's lives throughout history, and it continues to stay important and relevant today.

While this book details the equipment, ingredients, methods, and formulas needed to make excellent breads, this chapter serves as an introduction to the *concept* of artisan bread—what that term means, how it has evolved, and how it relates to the breads we make today.

IDENTIFYING THE TERM *ARTISAN*

What is artisan bread? The simple answer is that it is an excellent product made through careful labor, following proper methods. The complex answer has a little more meaning. At one point in history, bread baking was a basic skill that resulted in a simple product that served to provide nutrients. But the concept of bread baking has expanded to include dozens of categories of products, along with hundreds of types of breads and endless variations on those products. Equipment has been developed to streamline and perfect the process of making those breads. Bakers painstakingly study methods and techniques to ensure their full understanding of how to produce a proper artisan loaf. Bakeries have been created with the sole purpose of making and selling artisan bread.

The success in making truly artisan bread comes from understanding the entire process, from the twelve steps of bread baking (see page 84), to the equipment and ingredients needed (see pages 22–76), to the science behind each step in the process. When the baker fully understands the process, he or she can begin the extensive practicing of the techniques and the careful experimentation with formulas to produce the desired products. And once the desired products have been created, the artisan baker strives to achieve consistency in those products, ensuring that result through careful measurement and precision. When this has been accomplished, the artisan baker is confident that customers' expectations are being met.

To better understand how the production of artisan bread baking has come about, it is important to look at the history of bread baking.

A BRIEF HISTORY OF BREAD BAKING

Bread has been not only a major food source in countries around the world but also an important part of folklore, religion, and cultural tradition. That long, rich history dates from the first unleavened breads made in ancient Egypt to the modern artisan loaves made around the world today.

Early Bread Baking

Before the first bread could be made, ancient humans had to discover how to cultivate and harvest grains. This cultivation began in the Middle East, with early varieties of wheat, millet, and barley, some time between 9,000 and 10,000 years ago in what is termed the "Fertile Crescent," along the Nile, Tigris, and Euphrates rivers. The early farming centered primarily on two ancient forms of wheat: *einkorn* and *emmer*. These ancient varieties were different from modern commercial wheat in many ways, primarily in that they had double-bearded rows of kernels that produced a low yield and were difficult to separate from the husk.

At first, the ancient peoples simply boiled the harvested grains in water to create a porridge-like mixture. Over time, though, hybrid varieties of wheat developed, creating a stalk with more seeds that were easier to separate from the husk, thus drastically increasing the yield from each plant. These hybrid varieties also had a higher starch content, which made the grains easier to grind into flour. The primary wheat varietal used for bread baking today is *Triticum aestivum*, which is a descendant of that early emmer wheat variety.

The concept of the food we know as bread is said to have been developed in ancient Egypt about 5,000 years ago. There is archeological evidence that the ancient Egyptians understood the basic concept of fermentation, and they used that idea to produce both breads and alcoholic beverages. However, there is also some evidence that suggests earlier Stone Age humans had developed the concept of brewing beer from malted barley. Nevertheless, it appears that the ancient Egyptians can be credited with the first attempts at bread baking, as archeologists have uncovered remnants of their farming tools, grinding stones, and even their ovens. Added to this evidence are the depictions of bread baking that have been uncovered in Egyptian tombs.

The ancient Egyptians were able to grow vastly improved varieties of emmer wheat because their soils were made rich in nutrients by the annual flooding of the Nile River. Initially, the Egyptians ground the grains using a thick stone, a rectangular shape with a rounded top; this tool is now called a "quern." The Egyptian millers would place a second "roller" stone inside the quern, using it to press the grains, forming a meal. Over time, this tool was improved by fixing the stones together and attaching a turning handle. The handle was activated by the motion of animals, streamlining the process and increasing wheat yields. Though grain mills were eventually powered by steam and motors of various types, this basic principle of a rotary two-stoned mill remained unchanged until the advent of steel roller mills in the nineteenth century.

The Egyptians mixed coarsely milled flour with water and formed the dough into rustic grain cakes—the first flatbreads—which were cooked in the hot ashes near the open fire or atop a hot stone near the flames. The idea of risen bread appeared when the Egyptians decided to bake their bread inside pots, which provided constant, radiant heat. The pots were twin-halved baking pots with conical lids that could be heated separately from both sides. Eventually, this design was refined to a beehive-shaped oven in two tiers, with a firebox in the base and a baking platform, accessible by an arched opening. The beehive model is the basis of ovens used today.

The Role of Bread in Ancient Religion and Folklore

Bread is widely seen in ancient folklore and it has played a significant role in many religions. The ancient Egyptians shaped some bread loaves to represent a boar, and the loaves were dedicated to the fertility god Osiris. At that time also, Hebrew tribes considered the Egyptians unclean because they fermented their dough to make leavened bread. To the Hebrews, fermentation was impure, a form of corruption, and only unleavened breads were worthy of sacrifice to Jehovah.

Indeed, the ancient Greeks worshiped the source of their leavened bread as its own goddess, creating a myth reflecting the planting cycle of wheat. This is the story of the earth mother Demeter and her daughter Persephone (a seed princess). Persephone was abducted by Hades and taken to the Underworld for six months, able to return only in the spring each year, bringing the region into bloom and signifying the start of planting. The two also played a role in the nine-day celebration of bread, a feast that took place during the autumnal equinox. Romans later turned the Greek myth into their own, renaming Demeter as Ceres.

An obscure Jewish sect based in Palestine, the Essenes, preached the arrival of a Messiah who was said to be a bread god. According to revelations in the New Testament, Jesus was born in the house of bread (Bethlehem) as a human incarnation of the bread of heaven and the bread of God. Mary, mother of Jesus, was a descendant of earlier earth mothers whose sons were sacrificially killed in order to guarantee the continuance of life within the community. Early civilizations had seen bread as a symbol of life and fertility, but it was the Christians who elevated this symbol to the holy miracle known as the Eucharist, in which bread symbolizes the body of Christ.

In England, hot ash cakes were given to young girls as part of a tradition in which girls walked backwards to bed to dream of their future husbands. In many countries, bread is still placed in a bride's shoe and the shoe is attached to the getaway car. In Middle Eastern countries, dropped bread is a sign of bad luck, and the dropper must kiss it in hopes of reversing that luck.

Even today, bread plays an important role in traditions, folklore, and popular culture. Commonly, it is an icon of value: fertility, wealth, happiness, prosperity, or salvation. Consider the American slang of referring to money as "bread" or "dough," or commenting that a pregnant woman has a "bun in the oven."

Egypt became known as the "land of the bread eaters"; there is evidence that some pharaohs paid the workers who built the pyramids in bread and beer. Archeologists have even discovered remnants of a large-scale production bakery capable of producing 30,000 loaves a day; this bread works operated during the reign of Pharaoh Menkaure, whose pyramid was built nearly 4,500 years ago.

Bread Gains in Importance

While ancient Egypt set the stage for the cultivation of wheat, the Greeks and Romans continued down that path, and bread quickly assumed its important role in both cultures. However, as a primary food source and a household staple, bread also was central in political and trade issues.

For instance, the Greek Islands did not have proper terrain for growing wheat, and so the people relied on imported grain from Sicily. When the Roman Empire began dominating trade and expanding their routes into Asia, Rome began planting wheat on massive stretches of land along the northern coast of Africa. The Romans dispatched their engineers to build aqueducts and cisterns, thereby creating jobs and turning thousands of nomads into farmers. But while they were dominating the region's wheat production, the Romans made the mistake of keeping their wheat growers too far from their bread bakers, forcing the latter to endure high costs of importing their own grains.

In the first century A.D, Emperor Vespasian seized grain-trading fleets from the Flavian dynasty. In addition, he made all Roman bakers officials of the state, subjecting them to the political fortunes of Rome itself. At this time there were over 300 bakeries in Rome, many of which were run by Greeks, who had founded a guild for that purpose. Milling had been improved with the advent of horsepower rather than slaves, and with this improvement came greater and more readily available quantities of flour. This enabled the bakers to make a variety of breads ranging from white loaves for the upper classes to coarse loaves for the plebeians. The underclass was given a bread dole, which was notoriously and tyrannically controlled by the government. Over time, this took a huge political toll on the Roman Empire.

When invasions by the Vandals and Goths of the north caused Rome to fall in the fifth century, the focus shifted away from grains and toward livestock. The northern peoples were never an agricultural society like the Greeks or Romans, believing instead that the earth was a living thing and to alter it (for example, by plowing the land) was a sin. Livestock grazing fields quickly displaced the wheat fields, fermented barley replaced fermented grapes, and lard and butter took over the role played by olive oil.

In time, the Christians pushed out the Goths. As previously noted, bread was a vital part of Christian society, representing the body of Christ himself, so wheat production gained prominence once more. During the Middle Ages, castles were built like mini cities, fortified against each other and the surrounding countryside. This division intensified the gulf between bakers and farmers. Inside those cities, the bakers were the first craftsmen to form a guild, designating a pecking order of apprentice, journeyman, and master. At the same time, people began to distrust both millers and bakers, as their professions were crucial to those living within the city walls, so dependent as they were on that grain. With that increase in distrust came grain thefts, sales of inaccurate weights, and price hikes that only increased during times of famine. In some places, the process of grinding grain for flour was monitored closely to ensure that no individuals were secretly milling their own flour or baking their own bread.

During the Middle Ages, bakers expanded on the Roman tradition of baking different types of bread for different classes of people. In France, there were at least twenty varieties of bread, each named by rank: *pain de cour*, *pain de pape*, *pain de chevalier*, *pain de valet*, and so on. High-ranking nobles ate bread made from white wheat flours; commoners ate large round loaves named *pain de boulanger*, generally made from coarsely ground barley or rye flour. Similar divisions of bread were made in the church. After Pope Innocent III decreed the importance of the communion wafer in the early thirteenth century, Thomas Aquinas declared that the Eucharist must be made of only the finest wheat flour. The wafers also were unleavened to maintain purity. But this emphasis on white wheat flour only amplified the hierarchy of bread, maintaining the division between the bread of the rich and the bread of the poor.

The end of the Middle Ages marks the beginning of bread types being divided by region. While Germany maintained its feudal castle structure, Italy became a congregation of city-states and France developed as one major city with a dominant ruling class.

BREAD IN FRANCE

The importance of bread in France begins with the rise of the city of Paris in the ninth century, under the rule of Charlemagne. The twelfth century saw the first bakeries located inside the city walls; previously, ovens were kept outside the city in hopes of preventing fires. Every stage of baking was monitored by the royal court, including the type of flour to be used and the weight and price of the finished loaves. The crown even owned the ovens; it wasn't until the thirteenth century that bakers were allowed to rent ovens, thereby moving past the prior tradition of communal ovens. At court, the Grand Panetier served as purveyor and overseer of the king's bread; he also oversaw the setting and breaking down of the royal table at mealtimes.

By the end of the Middle Ages, the bakers had formed a coalition, designating rules and standards for bread baking. To become a master, bakers had to produce a "master piece" proving their skills, as well as pay annual dues. The masters presided over the valets, who were in charge of the apprentices (*geindres*). Masters shaped the bread, operated the ovens, and ran the business. Apprentices handled pest control and kneaded dough; the dough was often made in such large quantities that the kneading was done with the feet.

At this time in Paris, there were roughly thirty varieties of bread, differing in shape and ingredients. Popular breads included *pain de brode, pain coquille, pain balle, pain de ouleurs, pain rousset, pain de magne, pain de Chailly*, and *pain de Gonesse. Chailly* and *Gonesse* were both especially favored varieties—these towns were on the outskirts of Paris and were known for producing bread with an incredibly white crumb, likely due to the addition of alum.

The Renaissance marked major cultural shifts throughout Europe, led in large part by the Italians. By the seventeenth century, though, France had broken free of Italian dominance, and had begun to develop its own set of customs and cultural rules. This was prominently shown through their development of what would become modern cooking and dining.

The first major French cookbook, called *Le Cuisinier Francois,* was published in 1651, by La Varenne (a Marquis). This book was an important publication, but it focused on cuisine of the royal court. In 1654, a valet for Louis XIV, Nicholas de Bonnefons, wrote a cookbook more akin to home cooking—at least home cooking for the wealthy bourgeois. His *Les Delices de la Campagne* supplied the first important bread recipes, offered as a solution for the wealthy visiting their country homes who found the bread in the villages to be of poor quality. This cookbook included traditional sourdough breads and a bread similar to what would one day be known as brioche. The most noticeable characteristic of the recipes in de Bonnefons's book is the addition of beer yeast to leaven bread, rather than the existing natural process of fermentation. The beer yeast made the loaves rise more fully, creating a puffy appearance and giving a soft crumb. This style of bread making became increasingly popular under the reign of Louis XIV.

Thus, bread quickly became a symbol throughout France for the region's independence and rise to power. Incorporating bread prominently in their diet, rather than consuming grains in other forms like porridge, began to define the French identity. As time went on, even the poor were able to get their hands on white loaves, formerly reserved for royalty. Pale white bread with a thin crust became known throughout Europe as "French bread" and was designated with words like *pillow* or *pouf.* While some varieties of dark, dense bread were still made for the lower classes, a royal decree in 1670 permitted bakers to use beer yeast to produce the now infamous soft, white bread called *pain de luxe.* White bread was no longer a food reserved for the highest

Viennese Bread Traditions

The first noticeable Austrian contributions to bread baking came around the beginning of the nineteenth century, as the Austro-Hungarian Empire rose to power. Advances in milling in Hungary were producing high-quality flours, Viennese bakers had begun to use the poolish method (see page 124), and inventors were creating steam-injected ovens that produced products of superior quality. In addition, Austria was widely known for its excellent water (from the mountains) and its flavorful and unique natural yeasts. Hence, Vienna began taking the lead in the production of fine pastries, with an emphasis on intricately shaped yeast-raised pastries, some with flavorful fillings or toppings. These pastries were introduced to Paris in 1840, when an Austrian baron opened a shop selling this *pain viennois*. The 1873 World's Fair in Vienna solidified the city's reputation for fine baked goods.

One of the most well-known of these Austrian pastries was the croissant (see page 265). While breads had been made in crescent shapes for years to symbolize certain gods, fertility, and even the horns of bulls, croissants—made with flaky, yeast-raised puff pastry—are famously said to have come from Vienna. Indeed, the first croissants were made in Vienna in 1683, when the city was attacked by Turkish forces. It was the city's bakers, already awake and working on producing the day's bread, who alerted officials of the Turks' arrival. To celebrate the defeat of those invading forces, Austrian bakers made the pastries in a crescent shape, referencing the departed Ottoman Empire.

The flaky puff pastries filled with fruit or nut fillings we call Danishes (see page 254) were actually created by Viennese bakers who had been sent to Copenhagen during a shortage of skilled workers. The Danes referred to these pastries as "Vienna breads."

powers—the aristocracy and the church; it also represented personal health and wealth throughout France, for all its citizens.

The French Revolution in Bread

Bread is more than a symbol of the French Revolution—it's at the very core of that major political event. In the fifteenth century, millers and bakers joined forces in an attempt to deal with the ever-fluctuating prices and supplies of grains. As a way to settle the matter, the government enacted a law to control the price of bread. (This law was famously enforced until 1981.) However, the government could not control the price of European grain supplies, and the result was chaos. Both farmers and merchants took advantage of the fluctuations, often escalating their prices to exorbitant levels.

The French bread riots began as early 1725, as commoners grew increasingly angry about the famine caused by grain shortages and the rising grain prices. Indeed, by 1715, nearly one third of the French population had died of starvation. The bread riots resulted in the Flour War of 1774, expanding to involve countries like Austria,

England, and Prussia. Through all of this, the art of bread baking in France was being elevated to incredible heights, the products admired and renowned by countries around the world.

In 1767, Paul Jacques Malouin published a series of illustrations detailing the layout and equipment of a bakery, along with descriptions of various types of breads being produced at the time. His detailed accounts explained much about bread's role in French culture and history. Malouin divided the social politics of bread into three divisions:

- *Pain du Blanc* (*pâte molle*): Bread made from white flour (very soft crumb)
- *Pain du Bis-Blanc* (*bâtarde*): Bread made from a mix of white and whole wheat flour (more dense crumb)
- *Pain du Bis* (*ferme*): Bread made from whole wheat flour (very dense crumb)

Bakers preferred to make *pâte molle*, not only because of its greater desirability but also because it used more water and less flour—meaning it was less expensive to make but could be sold at a higher price. The Paris ordinance of 1635 demanded that bakers show only bread for the commoners in the windows of their shops, while *pain du blanc* loaves were to be held behind the counters. That made these breads only increase in popularity, their most desired trait being their thin, crisp crust. So, to meet customer demands, bakers began rolling the dough into longer loaves to increase the surface area of the crust. In his book, Malouin also discussed the reliance of the French on wheat flours, imploring the government to research alternative flours. In an effort to ward off famine, in 1778, the crown funded the Academy of Baking in Paris, headed by Antoine Augustin Parmentier, to research such alternative flours.

However, economic problems only worsened; by 1789, wheat flour had become so scarce that bread doubled in price. The French population first turned their anger toward the bakers, who pointed fingers at the millers, claiming they were unable to purchase adequate quantities of flour. Rumors began to circulate of a massive conspiracy headed by the Minister of Finance, who was thought to be working along with farmers, millers, and bakers to collectively raise the price of grain. The unrest reached its limit on July 14, 1789, when commoners stormed the Bastille in a belief that there was grain surplus stored inside.

A drought in 1789 only worsened the wheat shortage, and in October of that year the commoners also stormed Versailles, bringing the French king, queen, and dauphin captive to Paris. Over the next few years, France suffered war, poor harvests, and high levels of inflation. By 1793, the Paris Commune decreed that only *pain du bis* (whole wheat) would be sold in bakeries, in an effort to produce *pain de'eglaite*, or "bread equality." An allotment of 1½ pounds of bread was issued to workers and heads of family, and 1 pound to all other citizens. However, just a few years later, *pain de blanc* became the new bread representing "equality." It was believed in post-Revolutionary France that this bread, which had become iconic across Europe, was good enough for every citizen of France.

Revolutionizing the Technology of Bread

The nineteenth century marked a series of major revolutions in the processes of growing, harvesting, and milling grains. Advances in soil fertility were discovered by France's Louis Pasteur and Germany's Justus von Liebig, reducing the reliance of farmers on crop rotation. Austria's Gregor Mendel began studying the manipulation of crops by cross-breeding grains to improve desirable characteristics. Then, Russian scientists used these concepts to develop a strain of wheat that could survive their harsh winters. In North America, scientists focused on breeding wheat that could grow on the prairies of Canada and the United States.

And advances weren't limited to the science of farming. American Cyrus McCormick invented new harvesting machinery: a mechanical reaper and binder. These inventions, combined with massive crops being grown in the United States, as well as new railroad and steamship technology, enabled huge quantities of grain to be shipped across America and to Europe as well.

Swiss inventors began designing a new method of milling in 1834 that utilized rollers. This technique, called "the Hungarian system," used rollers rather than rotating stones to grind the wheat. The system moved beyond prior milling techniques, which had simply crushed the wheat kernels. Now, the bran and germ were separated from the endosperm. Successive sets of rollers would then extract more and more of the bran, which was removed by air drafts or by sifting through the screens. (Previously, the flour underwent a time-consuming process of being bolted by hand through silk gauze to remove the bran.)

These milling advances also increased the shelf life of the resulting flour. Because the germ was removed, the oil content was lowered, making the flour less prone to rancidity. Around the same time, millers started bleaching the flour to make it whiter in a shorter time period (flour naturally becomes white as it is allowed to mature).

In the bakeshop, technology began to streamline the process of making and baking bread, as well. Electric and gas-powered mixers were in use toward the end of the nineteenth century. But even before this, ovens powered by gas, oil, and electricity had begun replacing wood-fired ovens. Likewise, refrigeration changed the bakers' schedule, altering their production period and not limiting them to working overnight. The introduction of compressed yeast gave bakers increased flexibility in creating the bread formulas and timing how dough was proofed.

The year 1800 marked the first use of a preferment other than the traditional sourdough method. A Polish baker developed the concept of a "sponge," known later in France as *poolish* (see page 124). This method streamlined the process of making bread, as the poolish was essentially a piece of prefermented dough without beer yeast. This produced a highly desirable end product and was less sour than the sourdough leavener.

France initially did not adopt any industrialization of bread baking, but in the 1960s, technology won, and the process swerved away from the artisanal methods that had made French bread famous. Very few bakers held onto the traditional methods. However, one baker, Pierre Poilane and his apprentice son Lionel, maintained those standards of high-quality bread. In the 1970s, Lionel and his brother Max spearheaded a movement to bring artisanal bread-baking methods back to prominence throughout France.

French Regional Breads. From the left: Epi Baguettes (page 134), Tricouronne (page 139), Brioche à Tête (page 271), Pain au Levain (page 167), Fougasse aux Olives (page 313), Gugelhopf (page 222).

French Bread Shapes and French Bread Today

The baguette is France's iconic loaf, but it wasn't named as such until the 1930s. Long before that, these skinny loaves had been popular throughout Paris. Additionally, they were ideally suited to the technological advances of the day (commonly using poolish and mechanical kneaders for mixing). Their long shape allowed the loaf to rise and bake quickly—but it also staled quickly. This meant that baguettes needed to be baked, purchased, and eaten fresh; in Paris, this meant three times a day.

But there are dozens of French breads beyond the infamous baguette. The French notoriously classify many culinary and cultural traditions, and their breads are no exception. French breads can be named by any of the following methods:

- Type of leavening used
- Shape of the loaf
- Ingredients in the loaf

- Region where the bread originates
- Tradition, holiday, or event with which the bread is associated

Regional breads made a resurgence in the 1970s, many years after they had been banned during World War II (the government had standardized bread baking during the war to control costs and prevent fraudulent sales). While there are hundreds of varieties of bread in France, some examples of popular breads and regional varieties are:

- *Pain ordinaire*: A plain, white loaf, generally using no preferment.
- *Pain complet*: A whole wheat version of *pain ordinaire*.
- *Pain de mie*: The sandwich bread of France, made with milk, butter or oil, and sugar and baked in a Pullman loaf pan, with a tight crumb for easy slicing.
- *Pain de siegle*: French rye bread, made with at least 66 percent rye flour (otherwise it is *pain au siegle*), with a dark crust, commonly eaten with oysters. The addition of raisins is common, which makes it *pain de siegle rustique*.
- *Pain de campagne*: Bread made with both white and rye flours, leavened with sourdough starter, shaped into a large, round loaf (4 to 6 lb/1.81 to 2.72 kg), with a dark brown crust.
- *Pâte morte:* Also known as "dead dough," a plain dough enriched with fat but without the use of leaveners. It is used to make decorations on loaves for garnish or display (see photo on page 378).
- *Pain de metiel*: Made from wheat flour mixed with buckwheat, barley, or rye flour. Traditionally during the use of communal ovens, housewives would dust their initials on top of the loaf in flour, and some bakeries still continue this practice today.
- *Pain aux cereales*: Whole-grain bread made with sesame seeds, linseed, and malt.
- *Pain grenoblois:* Regional bread from Grenoble, made with wheat and rye flour, as well as sugar and milk.
- *Pain bordelais:* Regional bread from Bordeaux, made with wheat, rye, and whole wheat flours.
- *Pain normand:* Regional bread from Normandy made with cooked apples and apple cider.
- *Pains de provence*: Any number of regional breads from Provence.
- *Fougasse:* A regional bread from Provence, shaped into a flat loaf with open pieces in the center, like a ladder. These loaves are often stuffed with olives, bacon, anchovies, or nuts and are flavored with olive oil and herbs.

- *Pissaladière*: A pizza-like bread from Nice and Marseille, topped with anchovies, caramelized onions, and olives.

While bread is consumed in lesser quantities in France today (at one time, a typical Frenchman ate 1 lb 12 oz/800 g of bread daily, whereas now the average is 5¼ oz/150 g), bread is still an important part of French culinary culture and is expected to be served at every meal. The baguette maintains popularity for this reason—it can be served alongside any meal, regardless of the time of day, formality, or ingredients being used.

BREAD IN ITALY

Italy's culture is greatly centered on food and cooking, and bread plays an important role in that. The country's location (largely surrounded by sea) made it vulnerable to invasions. Italy was invaded first by the Etruscans, then the Greeks, then the Saracens,

Italian regional breads. From the left: Pane Siciliano (page 186), Pane alle Olive (page 165), Focaccia (page 324), Francese Stiratto, Panettone (page 249), Semolina Bread (page 182).

and so on. Italy became a melting pot of ethnic populations and cultures, as well as a gateway between the East and West. This position is only amplified by Italy's rich history, including the rise and dominance of Roman Empire, as well as the importance of the church to Italy and the rest of Europe. While Italian breads are widely known as being solely Italian, the inspiration for many of these breads is undeniably countries in the Middle East and Asia Minor.

From the fall of Rome until the rise of General Garibaldi in the middle of the nineteenth century, Italy was a collection of city-states. Even today, the majority of the Italian population lives in villages, hill towns, and small cities—often called "peasant cities"—in one of the twelve distinct Italian regions. Unlike France, where everything is centered on a single major city, there are six large cities in Italy: Milan and Turin in the north, Florence and Rome in the center, and Naples and Palermo in the south. Italy was slower to industrialize than either France or Germany, and until the middle of the twentieth century, power was held by a small group of landowners whose income was based on rent paid rather than crops produced on their land. This lasted until the Land Reform Acts of 1950, meaning that the traditions local to specific regions have been

Italian Special-Occasion Breads

Special breads are made throughout Italy to celebrate a variety of occasions, from agricultural celebrations, to religious events, to annual festivals. Many of these breads are merely decorative and are made from *granoduro*, or hard wheat, and are barely leavened. The result is a stiff dough that can be easily shaped into a variety of styles. Some of the most common special-occasion breads celebrate the following:

- San Guiseppe, the patron saint of pastry cooks, is given an annual festival on March 19. At home, an altar is set up dedicated to the saint, including flowers and three kinds of bread, which are decorative and not meant to be eaten. The first tier of the altar is made up of three huge rounds of bread, sometimes weighing as much as 20 lb/9.08 kg, representing the holy family. The altar is then decorated with bread shaped to represent Joseph's staff, a date palm for Mary, and a wreath for Jesus. Above the altar are hung bay leaves and citrus, along with tiny breads shaped as moons and stars. Bread shaped into Mary's slippers is placed in front of the altar—this symbol is a treasured one in Italy, making the Mother Mary seem intimate and domestic.
- Christmas in Rome is a highly celebrated time, and in early Christian times, worshipers offered focaccia bread, as well as breads shaped into discs to represent the sun, to altars of Christ. *Panedi Natale* was a popular Christmas bread in the Middle Ages. Enriched with fat, sugar, eggs, nuts, and dried fruit, it served as a symbol of plenitude. In Florence, breads were flavored with walnuts, pine nuts, honey, figs, and dates. In Emilia-Romagna, bakers added raisins, black pepper, and candied pumpkin. Traditionally, each member of the family would take a bite of the first three slices to ensure good luck in the coming year. In Ferrara, bakers present *pan pepato,* a chocolate-covered loaf, to their regular clients as a Christmas gift.

maintained. These traditions include religious and agricultural festivals, each with its own special foods, including bread.

Italy's vastly diverse geographical regions have helped dictate what kinds of bread are produced in each locale. Eighty percent of Italy is mountainous, with the Apennines running from the Alps in the north to the tip of Calabria, separating the eastern and western parts of Italy, giving many spots access to mountains, valley, and sea. Temperatures are regulated by the Mediterranean Sea. Generally, soft wheat is grown in the north and hard wheat in the south, with other important crops like olives, tomatoes, and grapes grown throughout the country.

Many breads in Italy use a preferment called a *biga* (see page 125). Firmer than a poolish, the biga serves the same functions of shortening the fermentation time and increasing the flavor. To best understand the breads of Italy, it is ideal to review them based on the regions from where they originate.

Tuscany's Breads

One of Italy's true epicurean epicenters, Tuscany is home to several of Italy's most famous varieties of bread, *pane toscano*. Traditionally, these breads were made without salt—possibly originally because of high prices on salt and taxes on it. But the more commonly believed reason was taste. *Pane toscano* was traditionally paired with very flavorful foods. The prosciutto in Tuscany, for example, is one of the saltiest varieties in Italy; and the olive oil produced in Lucca is one of the most flavorful and fragrant. Tuscany so values pure flavors that *pane toscano* is used to highlight the flavors of the food that it is served with. One of the most popular dishes of the region is *la fettuna*, a slice of Tuscan bread slathered with oil and warmed in the oven.

Other breads in the region include *pane toscanoscuro*, a whole wheat version of *pane toscano* scored with a checkerboard pattern on top. *Panediterni*, a regional bread from Terni, is another saltless bread, lighter in color and texture, with an open crumb structure. *Pane schiacciata*, or "crushed bread," is a traditional Tuscan hearth bread baked on an earthenware slab in a wood-fired oven. The dough is stretched thin, pricked with a fork, and then covered with a lid that is piled with hot embers.

Northern and Central Italy's Breads

The most well-known bread of northern Italy is *ciabatta*, a thin loaf made from a very wet dough with a crunchy crust and an open, chewy crumb structure. The word *ciabatta* means "slipper" or "old shoe" and is somewhat descriptive of the thin, wonky shape. Near the Swiss and French borders, *pane franchese*, made with a sour starter and a mixture of white and whole wheat flours, is popular. *Pan di como* is another version of this, made with only white wheat flour, shaped and risen in baskets, and leavened so that the dough forms high domes, with an open, honeycomb-like interior.

Also popular in this region is a dense rye bread, called *pain de segale*, which is made with a sponge preferment to give the bread a moist interior texture. Often, pancetta is added to the dough as well. Milan is home to *franchese stirato*, meaning "French string," a thin crusty loaf. *Filone* is similar to a baguette, made with a sourdough starter and with a crisp exterior crust and a tender interior crumb. *Pane di noci*, or walnut bread, uses a mixture of white and whole wheat flours along with chopped walnuts. *Pane casereccio* (Caesar's bread) has a very fine crumb and crispy crust.

Pain di mais, or corn bread, is the celebrated bread of the northern corn-growing region in the Po Valley, which also produces the popular dish polenta. Using finely ground corn flour mixed with wheat flour, the bread is shaped in large, rustic rounds and has a golden crumb color. *Pane di patate,* or potato bread, is made in the neighboring region of Valle d'Aosta, using mashed potatoes, oil, garlic, and parsley.

Small varieties of bread and thin, crisp crackers and breadsticks are also very popular in the northern regions of Italy. Some varieties include *panini*, a thinner bread that is halved and used for making sandwiches. *Biova* are soft rolls with a fluffy texture. *Coppiete* are starfish-shaped rolls from the Emilia-Romagna region. *Michetta* from Milan and *rosette veneziane* from Venice are rosette-shaped rolls that require special stamps to form the small rounds. These rolls are sometimes rolled in sugar, have almost no interior crumb structure, and are made up almost entirely of crisp crust. *Semelles*, originating from Florence and Rome, are another roll made in this fashion.

Grissini torinesi are the famous thin breadsticks of Turin. Napoleon called them "little batons" and was said to snack on them voraciously. They were apparently created by a baker working in the court in the seventeenth century to serve to a prince suffering from indigestion. It was presumed that these small, crunchy sticks were easier on his stomach than dense, heavy bread. *Grissini* are usually brushed with oil and topped with cheese, poppy seeds, or sesame seeds.

Pane all'olio is a flavorful dough enriched with both olive oil and lard, shaped into rolls. The rolls are then put together in a ring to make a wreath shape and sprinkled with coarse salt. *Pan marino*, or rosemary bread, is an aromatic round loaf garnished generously with salt.

Southern Italy's Breads

Puglia is one of the most popular bread-producing regions of southern Italy. *Friselle* is one variety, shaped into a flat, crisp bread with a hole in the middle. This bread is

Panettone

Milan's prized Christmas bread, *panettone* has become popular throughout the world, enjoyed year-round. The bread is baked in a high-sided, rounded mold. The top crust is a deep brown, while the interior is fluffy, light, and studded with dried fruit and nuts. The characteristic shape was the result of a cylindrical mold used by the Angelo Motta Company in the 1920s. The loaves were so popular that a competitor, Giacchino Allemagna, who had previously only made large (10 lb/4.54 kg) loaves, began churning out smaller varieties (as small as 1 lb/454 g), so that anyone could have a *pannetone,* now a revered status symbol, on their Christmas table.

sometimes made as small, individual pieces (like bagels) and other times made into large loaves. The dough is baked as a whole, then split in half and returned to the oven, where it bakes until it is very crisp. *Pane pugliese* is a large loaf (up to 4 lb 6½ oz/2 kg), traditionally a commoner's bread found in the country. In times of famine, legend cursed anyone who wasted even a crumb of *pane pugliese*. It was said they were destined to spend as many years in purgatory as they wasted crumbs.

In the olive groves of southern Italy, *pane alle olive*, or olive bread, is made. After years of snacking on bread and olives, bakers began adding both green and black olives directly to the dough to form these loaves. Another version is *puccia*, which is based on whole wheat or durum rather than white flour.

Durum is a hard wheat that is ground into what is known as semolina flour. Lightly yellow in color, semolina flour is very high in gluten and can be ground coarsely or very fine. This is the same flour that is commonly used to make pasta dough throughout Italy. *Pane siciliano*, or Sicilian bread, mixes semolina and white wheat flour, along with olive oil. The loaf is commonly sprinkled with sesame seeds.

BREAD IN GERMANY AND EASTERN EUROPE

While artisan bread is often associated with France and Italy, eastern European countries have rich bread cultures that have often gone unnoticed. As in the other countries discussed, bread is the center of the food world, having a place at every meal. In Germany, supper is still called *abendbrot*, or "evening bread." In fact, it has been documented that Germans eat three times as much bread as do Americans, and more than most Europeans.

The countries that border the North and Baltic seas (Great Britain, Scandinavia, western Russia, Poland, Germany, Netherlands, and Belgium) were colonized by the

Rye Bread Gallery: 100% Rye Bread (page 349), Haferbrötchen (page 356), Landbrot mit Sauerkraut (page 359), St. Gallener's Brot (page 370), Rye and Sunflower Bread (page 373).

early Germanic peoples—a varied group united only by their lack of Latin speakers and their love of rye. Rye flour is to many parts of eastern Europe what wheat flour is to the western regions. While rye evolved later than wheat in the Middle East, by 3000 B.C.E. it was growing in southern Russia and in northern and central Europe. Rye grows well in cold climates, whereas wheat often dies.

Even before the Ostrogoths of the northeast and the Visigoths of the northwest conquered Rome, these tribes had split into twenty-five divergent groups, including the Teutons, Angles, Saxons, and Franks. Each tribe had its own dialect and was constantly growing and changing as its boundaries were shifting. The disunity of this region was amplified by its varied geographical differences; some places were bitterly cold and mountainous, while others were flooded and marshy.

Germany consisted largely of small farms surrounding small villages, which were overseen by local dukes. Despite Charlemagne's attempt to include Germany as part of the Holy Roman Empire in the ninth century, Germany was an autonomous state until

Bread in Eastern European Religion and Folklore

Beginning with early pagan myths and legends, Norse myths animated things found in nature, like thunder or fire. Breads shaped like animals, especially the wild boar, were often offered as a sacrifice to the powerful gods Thor, Odin, and Freyr. Easter and midsummer celebrations were marked with fires—peasants would run through the hot embers and dance around the fires, afterwards distributing loaves of bread and cakes.

Even after Christianity had successfully converted the populations in much of the north, farmers would enact ancient fertility rituals at their farms. In these rituals, grain was celebrated and seen as alive, usually personified as a woman: the old rye-mother. This character was also used to frighten small children away from the fields, where it was said that the rye-mother would strangle any children who trampled on her grains.

Fairy tales also mention the grain-mother, sometimes as a sacred being wielding positive powers, but most often as a demonic witch capable of evil or wrongdoing. In the original tale of Hansel and Gretel, the siblings are enticed to the gingerbread house by a "grain-witch."

Bismarck attempted to found a Prussian empire at the end of the nineteenth century. These separate villages had distinct cultural customs, including specific breads local to each region.

The Germans discriminated wildly between social classes. Aside from the usual divisions of nobility, clergy, burghers, and peasants, Germans also divided peasants into specific classes (landowner, hired help, foreman, chief maid, and so on). By the end of the Thirty Years War, large numbers of peasants had lost their land, and by the seventeenth century, nearly 60 percent of the population was gone—and the peasants that remained were incredibly poor until the industrial revolution brought changes before World War I.

Despite a distinct prominence of rye breads, wheat flour was often used to create enriched breads, decorative breads, or delicate products. Rye could not be ground as fine as wheat flour, and it was darker in color, making wheat flour ideal for breads where a certain appearance was desired. In many cases, wheat and rye flours were mixed in varying percentages to make a variety of products.

Varieties of Breads in Germany and Eastern Europe

There are hundreds of varieties of bread throughout eastern Europe, each of which has its own story, set of ingredients, and specific method of production. Some common varieties include:

- *Challah*: A lightly sweetened enriched bread originally made by the Ashkenazi Jews of southern Germany in the fifteenth century. This loaf gained popularity and eventually spread throughout Europe as a ritual bread for the Sabbath. The dough would be made on Thursday and would rise overnight, then be baked fresh for the Sabbath on Friday. The name comes from the Hebrew word for "portion," referencing the piece that would be offered to Jehovah. The bread can be made in a variety of shapes, but is most commonly braided.

- *Kugelhopf*: A buttery bread, similar to brioche, made with wheat flour and dried or candied fruit and citrus peel. The loaf was baked in a tall mold with a hole in the center. The mold is said to resemble a fez (the literal translation of *kugelhopf* is "Turk's head"). The bread is often baked in Alsace, the region near the German border, where it is consumed with Alsatian white wine.
- *Limpa*: A Swedish enriched bread made using rye flour and beer. A rustic loaf flavored with fennel, cardamom, and/or candied citrus zest.
- *Gebildbrote*: Translated as "picture bread," this is an elaborate genre of folk art in Germany dedicated to making bread sculptures and displays depicting religious traditions. *Gebackmodel* is a molded dough made for decorative display, and *Lebkuckenmodel* is a decorative gingerbread. The displays are so elaborate, they can look like detailed woodcuts.
- *Oster-Hefegeback*: Any one of a variety of decorative breads made to celebrate Easter in Germany, often shaped into nests, fish, or doves.
- *Speckkuchen*: A New Year's Day bread, stuffed with bacon and sometimes shaped into "good luck pigs," or *Neujahrs-Glucksschwienchen.*
- *Festbrot*: Special Christmas breads in Germany, made into shapes depicting different stories. *Wickelkindern* is one such bread—baked in a shoe shape with a child's head emerging from the toe. The bread is red and decorated with confectioners' sugar.
- *Fruchtebrot*: A special fruit bread baked at Christmas, sold only at the Christmas markets held in many German cities. *Stollen* is one variety. Another, *Dreikonigs Kuchen*, or "Three King's Bread," is shaped into a wreath and a trinket is hidden inside the dough.
- *Musli-Brotchen*: A hearty loaf developed in the Swiss Alpine region.
- *Haferbrot*: A loaf made with a large percentage of oats.
- *Leinsamenbrot*: A light rye bread with flaxseed.
- *Babka*: A Hungarian festive bread, traditionally filled with poppy seed filling.
- *Potato Bread*: A bread made from a base of mashed potatoes.

Pretzels

Pretzels originated from a tradition of shaped breads; it was common to make ring-shaped breads during the Roman Empire, as well as in Spain during the early days of Christianity, where dough bracelets would adorn the bodies of the dead before burial. Early pretzels were developed by monks somewhere between the fifth and seventh centuries. It is said that a monk was shaping scrap pieces of dough, and he twisted them into a shape to represent a person's arms crossed in prayer. At first, the popularity of pretzels was somewhat limited to the church: Monks used them to demonstrate Bible stories to children and also used them as rewards for their studies. It was thought that pretzels brought blessings and spiritual wholeness to those who ate them.

Eventually, pretzels made their way into wedding ceremonies, where couples would wish upon a pretzel, break it in half, and then eat it together to symbolize coming together as one. In fact, the twisted shape of the pretzel is the origin of the phrase "tying the knot."

Pretzels became increasingly popular in Germany, where children would tie pretzels on a string and wear them around their necks at the beginning of the new year to symbolize prosperity in the coming months. A pretzel topped with two hard-boiled eggs was hidden around Easter for children to find—likely the precursor to the modern Easter egg hunt.

By the Middle Ages, pretzels had become so popular that bakeries devoted solely to the craft of making pretzels opened. A giant pretzel would hang outside the door of the bakery or the baker would have a *Bretzen Back*, or "pretzel sign." Some bakers would blow a *Backerrufer*, or an "oxen horn," whenever hot pretzels had just come out of the oven.

There are forty-four named pretzel varieties, but they fall into three major categories: beer pretzels boiled in lye before baking and then sprinkled with salt, egg pretzels made with sweetened cookie dough and sprinkled with nuts, and soft pretzels made from bread doughs (much like bagels). Hard pretzels originated when the pretzels were dried to increase their shelf life.

- *Portugese Broa*: Made with a corn soaker similar to polenta, it is traditionally served with a classic Portuguese stew, caldo verde.
- *Vinchgauer*: An Austrian bread that originated in the late 1700s using spices that were prominently traded on the Spice Route and made its way to Austria, where it became famous.

2 EQUIPMENT

One of the truly unique things about bread baking as opposed to other baking specialties is that it needs very little equipment in general. However, the equipment that is required is specialized, often used exclusively for bread baking. In a bakeshop specializing in breads, it often makes sense to purchase some or all of this specialized equipment, which can truly streamline the bread-baking process. In a bakeshop or restaurant setting where bread baking is being incorporated into current production, some pieces of equipment may be less necessary. However, understanding the equipment and what each piece can accomplish will help the baker understand how to use these tools to create the best results possible.

For ease of reference, equipment in this section has been divided into two categories: large equipment and small equipment/hand tools.

LARGE EQUIPMENT

What is considered large equipment encompasses mainly appliances, including ovens, mixers, proof boxes, and so on. Because this type of equipment represents a large investment for a bread bakeshop, each product is detailed along with its advantages and disadvantages as they pertain to bread baking and production. This is intended to help the professional baker understand the differences in these pieces of equipment, but also to aid in choosing the right equipment if a purchase is being considered.

Ovens

Ovens are an integral piece of equipment in any bakeshop. Types and features of ovens can vary by manufacturer, but there are several types of ovens commonly used in bread baking. A bakeshop equipped with both a deck oven and a convection rack oven allows the most flexibility in terms of the range of products that can be produced, as well as the quantity of production.

DECK OVENS

A deck oven is so named for its multiple levels, stacked on top of one another and lined with a stone base. The stone is an excellent heat conductor, and it allows bread to be baked either in a pan atop the stone or, very commonly, directly on the stone's surface. Deck ovens are available with electric, gas, or steam tube heating elements. In addition to the basic heating component, deck ovens have a steam generator, which is ideal for bread baking (particularly for lean doughs), but can be used for other items as well. The steam generator injects steam into the oven at the beginning of baking, and then a release lever allows the steam to be vented out of the oven as desired during baking. Some deck ovens have steam generators for each deck, while others have one generator that produces the same amount of steam for all of the decks simultaneously.

Ovens with gas or steam tube heating elements present one of the primary considerations with these types of ovens, particularly important when purchasing. Deck ovens are large and ideal for large-scale bread baking and production. However, deck ovens with gas or steam tube heating elements are less flexible—each deck must be set to the same temperature. In large-scale production, this presents less of a problem, as each deck is likely to contain the same doughs because large quantities are being baked. However, the situation is less ideal for smaller batches or when trying to bake a variety of products. If more flexibility is desired, an electric deck oven with separate steam generators for each deck should be selected.

Advantages:

- Retain heat well and heat evenly and efficiently.
- Bread can be baked directly on the stone deck, which promotes better oven spring and a crisp crust.
- Last for a long time and require little maintenance.
- Can be used for a variety of products aside from bread baking—excellent for pastries and savory baking.
- In electric models, each deck can be set to a different temperature, increasing flexibility of use.

Disadvantages:

- Expensive to buy and install (especially the electric models).
- Labor-intensive to load and unload the oven.
- Gas and steam tube models are limited—all decks must be at the same temperature.

CONVECTION RACK OVENS

Convection rack ovens are large, metal-lined baking chambers with a rotating base. A fan inside the chamber circulates the hot air constantly and evenly. Convection rack ovens are available with electric or gas heating elements, though unlike deck ovens, there is little difference created by the heating component of the oven. Items are baked on pans that are loaded onto a rack—the rack is then wheeled inside the oven and secured at the base. This makes the oven incredibly easy to load and unload, though multiple racks are needed for production baking.

In addition, because these are large convection ovens, they are well suited to a wide array of products and can be used to bake many items other than bread. While the ovens have a lot of advantages in the form of ease, convenience, and flexibility, they don't produce the same results as a deck oven—namely the strong oven spring and crisp crust. But they do produce excellent rolls, as well as breads made with enriched doughs, such as brioche or challah.

Advantages:

- Less expensive than deck ovens.
- Suited for a wide array of baked goods.
- Take up very little floor space.
- Easy and fast to load and unload.

Disadvantages:

- Do not retain heat well.
- Difficult to achieve same crustiness and oven spring as in a deck oven.
- Require more maintenance over time because of the moving parts.

STANDARD CONVECTION OVENS

The deck and convection rack ovens are ideal for professional bread bakeshops. Some bakeries, hotels, restaurants, and the like may not choose to have an oven exclusively for bread baking. This could be because their bread items are not the primary products on their menu or simply because they do not produce large enough quantities. Bakeshops such as these may choose to use their standard convection oven when baking bread. Though these ovens are ideal for baking cakes, cookies, muffins, and pastries, they can also be used to bake bread, without some of the benefits of deck and rack ovens.

Advantages:

- Cost-effective.
- Suited for a wide array of baked goods.

- Take up very little floor space.
- Easy and fast to load and unload.
- Ease of operation.
- Excellent for baking enriched doughs, which require even baking and no crust development

Disadvantages:

- Crust development for lean dough would diminish.
- Finished appearance of crust, scoring, and sheen would diminish.
- Flavor would suffer because of the lack of crust development.
- Smaller size makes large production challenging.

WOOD-FIRED OVENS

Wood-fired ovens are, once again, increasingly popular in modern bakeries. These ovens utilize classic heat in the form of burning wood to bake a variety of products, and are especially excellent for yeast-raised breads. Wood-fired ovens can be expensive to install, require a decent amount of space, need proper ventilation, and call for special knowledge from the baker to understand and operate. Products baked in a wood-fired oven have excellent flavor, texture, and often a beautiful color from the very intense (though somewhat irregular) heat. Classic breads and rolls made from lean doughs bake excellently in a wood-fired oven—baguettes, ciabatta, semolina, and so on. Flatbreads like pita and naan, as well as pizzas, are also ideal, but any variety of baked goods can be baked in a wood-fired oven if the baker understands the fire and the heat produced by the oven. When the fire is built and stoked to the proper temperature, the baker must start with items that need very high heat (breads and rolls). As the fire dies down naturally, the baker can bake items that require less intense heat, such as Danish and croissant.

Advantages:

- Flavor.
- Heat radiating from the bottom of the oven enhances crust development, which also improves the look of the scoring and sheen.
- Better crust development results in better flavor.
- Oven is large enough to accommodate quantity production.

Disadvantages:

- Time to reheat is 2 to 4 hours if not used every day.
- Cleaning is time-consuming and difficult.

- Fuel takes up a lot of space, making storage challenging.
- Baking of different products must coincide with the diminishing heat of the oven.
- Cannot be reheated during the day.

Proofers and Retarders

Proofing is one of the twelve steps of bread baking discussed in Chapter 4. Traditionally, this step was done without the use of equipment; bread was just allowed to rise for a given period of time in a warm environment or at room temperature. Today, many professional bread bakers opt to purchase a proofer. The proof box provides a controlled warm, and ideally humid, environment for bread dough. With a proofer, the baker can adjust the temperature and humidity, thereby controlling the proofing process.

In addition to providing the optimal environment for bread to rise, most proofers also have the ability to serve as a retarder—basically, a controlled, cold environment—to stop the dough from rising. Most models can be programmed to act as a retarder, then switched to be a proofer at a specified time. This allows a baker to come into the bakeshop to perfectly proofed bread, ready to bake.

Proofers and retarders are available in a variety of styles. Some models are reach-in, with a rack where the bread is placed to proof. Others are roll-in, with the bread loaded onto a rack outside the proofer and rolled in. Some proofers have two units—one for proofing and one for retarding. Others are single units that perform both functions. In addition to the expense of purchasing and maintaining this piece of equipment, the baker should consider the energy cost of operating the proofer on a daily basis. Using a proofer/retarder can drastically streamline the bread-baking process—it gives the baker complete control over when a product is ready to bake, thereby allowing for a precise schedule to be followed.

Advantages:

- Allow for better organization and planning of the baking schedule.
- Provide a constant and consistent temperature regardless of weather or bakery conditions.
- Help organize shifts in bakery production.
- Provide a humid environment that keeps the product surface moist at all times.

Disadvantages:

- Take up floor space in the bakeshop.
- Expensive pieces of specialized equipment, plus incur energy costs to operate daily.

Work Surfaces

A wooden tabletop is the best work surface for bread baking. Marble and stainless steel, both commonly found in professional kitchens and bakeshops, are cold to the touch. Mixing and handling dough on one of these surfaces can cool the dough down, slowing the yeast growth and can affect the finished product. Wood, however, does not conduct heat, and is relatively easy to clean and care for over time, making it the ideal surface for a bread bakeshop.

Mixers

There are a variety of factors to consider when purchasing a mixer for a bread bakeshop. One of the most important is size—most mixers are available in several sizes, generally designated by bowl capacity. Remember that mixing a small amount of dough in a large mixer can cause it to be mixed improperly. In contrast, overloading a mixer or filling it very close to capacity can also lead to uneven mixing and can cause wear of the motor, shortening its life span. Having mixers of varying sizes can be a solution to this problem, but cost can be a factor.

SPIRAL MIXERS

Spiral mixers are specifically designed for mixing bread dough, making them an ideal tool in a bread bakeshop. Available in a variety of sizes (generally ranging from 5- to 120-quart capacity, up to 400-quart bowls), these powerful mixers have two moving parts. In addition to a spinning spiral-shaped hook, the bowl automatically rotates. The mixers generally have three speeds. The hook spins twice as fast as the bowl, and some models allow for the bowl to spin both clockwise and counterclockwise, as needed. The actions of the spiral mixer imitate the stretching and folding that is achieved when dough is mixed by hand.

The main advantage of spiral mixers is their power, along with the spinning bowl, which encourages even mixing. However, because they mix so well and so efficiently, care must be taken to prevent overmixing. Careful monitoring during the initial mixing of a formula can help determine the proper mixing speeds and times (keep in mind that correct mixing reflects the skill of an experienced baker). Like many larger mixers, these mixers have a metal guard that covers the bowl during mixing and that guard must be in place before the mixing function can begin. The guard prevents the operator of the machine from touching the spiral attachment when the mixer is in use and getting injured.

Advantages:

- Fast and efficient dough mixing—better than other mixers for bread dough.
- Work well with both small and large quantities of dough.
- Equally effective with a variety of dough consistencies (wet, dry, sticky, etc).

Disadvantages:

- Care must be taken to prevent overmixing.
- Specifically designed for dough production, making it less flexible for a bakeshop producing items other than breads.

VERTICAL PLANETARY MIXERS

These multipurpose mixers are the common model found in most bakeshops; they are composed of a motor housed inside a metal body, a stationary removable bowl, and several removable attachments. Available in a variety of sizes (bowls generally ranging from 5 to 120 quarts, up to 400 quarts), these mixers have three or four speeds. There are three available attachments, which have planetary movement around the bowl. This type of mixing is ideal for certain purposes, like enriched doughs. The dough hook is a large, curved metal hook specifically designed for mixing bread dough. The attachment imitates kneading, as it pushes the dough against the base and sides of the mixing bowl. The paddle attachment is designed for mixing doughs and batters using the creaming and blending method (ideal for cookies, cakes, and other liquid batters). The whip attachment is designed for incorporating air into the ingredients in the mixing bowl, and is generally used for the foaming methods.

Advantages:

- The different attachments make these mixers versatile in the bakeshop.
- Well suited for smaller quantities.
- Planetary action is more efficient for certain enriched doughs.

Disadvantages:

- Planetary movement is not efficient for developing most bread doughs.
- Due to the mixing motion, these mixers have a tendency to increase the temperature of the dough quickly.

OBLIQUE MIXERS

These mixers, also known as fork mixers, are similar in construction to spiral mixers, except that the mixing attachment is a fork rather than a spiral. These mixers are used exclusively for bread baking, and, like spiral mixers, they minimize the friction in the mixer by working the dough gently throughout the mixing.

Advantages:

- Ideal environment for mixing bread doughs.
- Fast and efficient dough mixing with minimal mixer friction.

Disadvantages:

- Specifically designed for dough production, making them less flexible for a bake-shop producing items other than breads.

Water Chillers and Meters

This piece of specialized equipment is ideal for the breads bakeshop. Consisting of a tank fitted with a cooling element, it is designed to measure a specific amount of water at just the right temperature. Water can be cooled to the desired temperature with the use of a meter. In a large production bakeshop, this piece of equipment can save a lot of time during the scaling and mixing processes, especially if a large variety of doughs are being mixed.

Advantages:

- Cold water is available at all times without wasting time, using ice, or taking up refrigerator space.
- Ensure accurate temperatures and measurements.

Disadvantages:

- Expensive and a single-use product.
- Take up floor space in the bakeshop.
- Some water will be wasted in order to achieve the correct temperature.

Dough Dividers and Molders

A dough divider is a machine that can quickly and accurately divide and round off dough into desired weights. It can be used to divide dough into small quantities (for rolls) or larger quantities (for loaves). Dough dividers are available in several styles: manual, hydraulic, and volumetric. In addition, there is the divider rounder, which not only divides the dough but also shapes the pieces. Lastly, a molder cuts the dough into baguette lengths and shapes the pieces before baking.

MANUAL DUTCHESS DOUGH DIVIDERS

The baker places a weighed portion of dough onto a plate in the manual dough divider, and the dough then is put into the machine under a cutting device. The weight of the dough on the plate determines the number of pieces of dough by multiplying the desired portion of dough for each piece by the number of cutting sections on the machine. The cutters are lowered into the dough by exerting pressure on the handle, thereby cutting the dough into equal portions.

Advantages:

- Less expensive than other dividers.
- More compact than other models, taking up less floor space in the bakeshop.

Disadvantages:

- Specifically designed for making rolls, with less flexibility for dividing larger quantities of dough.
- The process is slow when doing a large quantity of rolls.
- Dough still needs additional shaping by hand before baking.

HYDRAULIC DIVIDERS

The baker places a weighed piece of dough into a square or round chamber in this type of divider, which is then closed for the cutting process. The weight of the dough placed in the machine determines the number of pieces of dough by multiplying the desired portion of dough for each piece by the number of cutting sections on the machine. When the lid is closed, the bottom of the container is pushed up by a hydraulic pump until the dough fills the chamber. Then the cutters are pushed up from the bottom, through the dough, to the lid. The cutters drop, the lid opens, and the dough is pushed to the top of the chamber for easy removal.

Advantages:

- Work gently with the dough, causing less ripping and tearing.
- Well suited for both wetter and drier doughs.
- Fast and efficient.

Disadvantages:

- Pressure may increase the strength of the dough.
- If the dough is not spread properly, the pieces will be irregular and will not weigh the same.

VOLUMETRIC DIVIDERS

The baker places the dough in a hopper, from which it is sucked into a cylindrical chamber, which is used to measure the portion of dough. That is, the dough is measured by volume and can be adjusted by adjusting the size of the cylinder. Some hoppers are open, which allows the dough to be pulled into the cylinder by a piston. Others are closed, creating a pressurized system that forces the dough into the cylinder. The measured amount of dough is then cut, and the portions are discharged on a conveyer belt.

Advantages:

- Work at a very high speed and divides dough very quickly.
- Large amounts of dough can go directly into the hopper, streamlining the dividing process.
- Well suited for an automated production line.

Disadvantages:

- Dough structure can be damaged as dough moves into the cylinder.
- Larger margin of error when measuring gassy doughs by volume.
- Weights may vary with volume, making final divided dough less accurate.
- Some machines have difficulty dividing very stiff or very soft doughs.

DIVIDER ROUNDERS

These machines are automatic or semi-automatic dividers that not only divide the dough for the rolls but also round them. The dividing and portioning action is the same as in the manual dough divider, but the rounder model also features a lever that initiates an automated rounding motion after cutting to form evenly rounded rolls.

Advantages:

- One piece of equipment completes two steps.
- Accurate and consistent product.
- Time efficient.

Disadvantages:

- Cannot accommodate larger quantities of doughs or divide for loaves.
- May not work well with gassy doughs.

BAGUETTE MOLDERS

These molders mechanically imitate the process of hand shaping the baguettes, but increase the speed, which therefore improves the consistency of the shaping. The machine consists of two or three sets of rollers that flatten the dough, curl it into a cylinder, and elongate it into the final shape. The molder has settings that can be adjusted to alter the size and tightness of the final loaf. While this equipment is specialized and particular to just baguettes, it enables a single person to do the work of many bakers, and can strongly reduce production time by streamlining the shaping process.

Advantages:

- If used properly, the results are the same or better than hand shaping.
- Relatively compact, they do not take up too much space on the bakeshop floor.
- Fast and efficient.
- Adjustable to a variety of sizes.

Disadvantages:

- If not adjusted properly, the structure of the dough can be damaged.
- Difficult to use with very hydrated doughs.

Reversible Dough Sheeters

These machines roll out dough to the desired thickness and size. They function by passing the dough back and forth through an adjustable set of rollers. Sheeters are available as both tabletop and floor models. Some are fully automated, and require little manual assistance. Many have accessories that can perform other functions, such as cutting or shaping items such as croissants.

Advantages:

- Save time and labor.
- Apply even pressure to the dough, ensuring it's rolled out evenly.
- Can be used for a variety of doughs, for a variety of products.

Disadvantages:

- Expensive.
- Take up a large amount of space in the bakeshop.

Oven Loaders

Oven loaders are used to transfer shaped dough to the oven for baking. The loader is made up of a metal frame covered with a fabric stretcher. The loaves are placed directly onto the fabric, and then the entire machine is pushed into the oven. After it is inside the oven, the handle is pulled, which unloads the loaves directly onto the base of the oven.

Advantages:

- Quick and efficient.
- Load proofed loaves without disturbing the structure or volume.
- Allow for organized spacing of bread inside the oven.
- Allow maximum use of oven space.

Disadvantages:

- Space is needed to hold the loader for installation, storage, and loading.

Speed Racks and Cooling Racks

Indispensible tools in the bakeshop, speed racks and cooling racks are sets of multilevel racks that can hold sheet trays, wooden trays with couches, or finished loaves for cooling. Speed racks are generally tall, thinner metal racks the proper width and length for holding a sheet tray. Specially made racks can also be used in the proofing process (wheeled into a proof box or retarder) and in baking (wheeled into a convection rack oven). Cooling racks can be metal, wood, or plastic and have shelving to hold items in various shapes and sizes (namely baked breads and rolls). The shelving has slats to provide ventilation around the bread as it cools. Both are on wheels to make them easy to maneuver around the bakeshop.

SMALL TOOLS AND HAND TOOLS

This category of tools includes many of the most important tools in bread production, though it omits one very important tool. A baker's hands are by far the most crucial tool. They can facilitate the use of tools, but they also act as tools of their own—as mixer, scraper, even thermometer. But most important, hands allow the baker to feel the dough, which leads to a better understanding of every aspect of production. As with the larger tools described earlier, these small tools streamline the process at various stages of production.

Scales

Baking, and especially bread baking, requires precise measurement. Proper measurements ensure accuracy, as well as consistency of quality and quantity. For this reason,

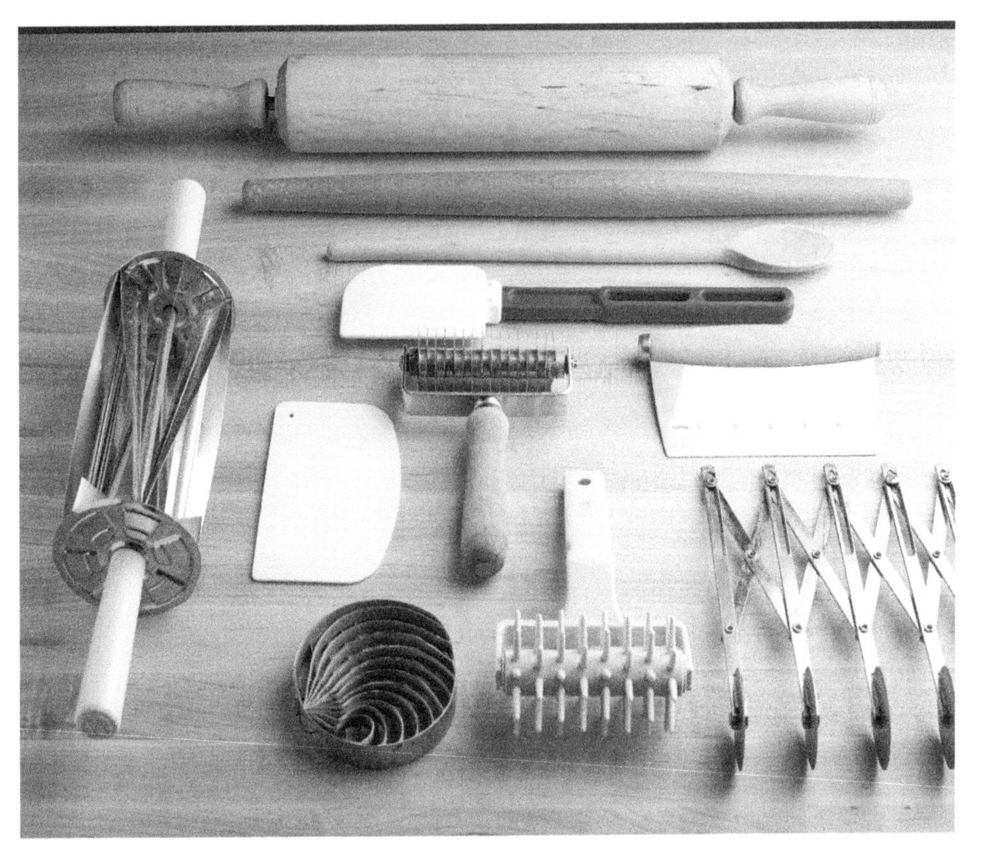

From the top left: rolling pin, French rolling pin, croissant cutter, heat-resistant rubber spatula, wooden spoon, bowl scraper, lattice cutter, concha cutter, dough docker, bench knife, expandable pastry wheel.

and because of the larger quantities of ingredients most often used in commercial bread baking, scales are the most desirable form of measuring ingredients before mixing.

DIGITAL SCALES

Digital scales have a platform set on an electric base that houses a digital display. These scales come in a variety of sizes, but smaller scales can accommodate only up to a certain weight, though they tend to be more accurate judging smaller quantities. Larger scales can weigh larger amounts, but they are less sensitive and therefore can be less accurate about smaller quantities. To tare a digital scale, simply press the "tare" or "zero" button until it reads zero. Most digital scales switch between U.S. and metric weight systems with the touch of a button.

SPRING SCALES

Spring scales have a platform set on top of a pedestal that contains a spring mechanism for weighing, and there's a dial indicator on the front. To tare a spring scale, place the container for measuring the ingredients on the scale and turn the dial so that the pointer aligns with the zero.

BEAM BALANCE OR BAKER'S SCALES

These traditional scales have two platforms attached on either end of a beam. The point where the beam and base meet is called the fulcrum. At the front of the scale, a weight hangs from a bar available in metric and U.S. measurements. To tare the scale, set the container used for weighing the ingredient on one of the platforms. Reset the scale so that both sides are level, either by manipulating the weight on the front of the scale or by adding a counterweight to the other side of the scale. Ingredients can be weighed using the same method—either manipulating the front weight until the two platforms are even or adding counterweights. These scales are very durable and the use is quick and efficient.

Thermometers

Thermometers are important at various stages of bread baking—for example, taking the temperature of the ingredients before mixing, taking the temperature of a dough after mixing and during proofing, even taking the temperature of a baked loaf to ensure doneness. It's also a good idea to hang thermometers in the bakeshop to make sure there is an accurate reading of the room temperature, which is needed to calculate DDT (desired dough temperature; see page 82).

INSTANT-READ THERMOMETERS

Instant-read thermometers are available with both dial and digital readouts. Digital thermometers typically measure a wider range of temperatures and also measure more accurately. They are also more accurate when measuring shallow liquids.

STEM THERMOMETERS

Stem thermometers are ideal for measuring the internal temperature of products, even during baking. They have a long stem with a digital or dial head that indicates the temperature.

PROBE THERMOMETERS

Similar to stem thermometers, probe thermometers have a much shorter cord that attaches the metal probe to the digital readout. Some also have an alarm setting to indicate that a specific temperature has been reached.

Dough Storage Containers

To save counter or bench space during production, it is ideal to have proper containers for storing dough during bulk fermentation. The best material is plastic because it is lightweight and easy to clean. It is important that the container be large enough to provide room for the dough to rise throughout fermentation. A lid will prevent the dough from being exposed directly to air, though other types of coverings can be used if the container does not come with a lid (see page 39).

Bench and Bowl Scrapers

These multi-use scrapers are indispensible in the bread bakeshop. Bench scrapers, or bench knives, have a wide metal blade attached to a handle, and the handle can be made of plastic, wood, or metal. The edges are not sharp like a knife, but they can cut easily through dough. In addition, bench knives and scrapers are ideal for scraping and lifting, and they aid in dough folding. Bowl scrapers, or plastic scrapers, are flexible plastic scrapers used to scrape clean the mixing bowls or dough storage containers, leaving no waste behind. Some are rounded on one side. Both bench and bowl scrapers can also be used to help clean equipment and work surfaces in the bakeshop.

Bench and Pastry Brushes

Brushes are used in the bakeshop to remove excess flour. Bench brushes are large brushes fitted to a plastic or wooden handle. Their larger size makes them ideal for brushing flour off of equipment and work surfaces. Pastry brushes are small brushes fitted to a plastic or wooden handle. Their smaller size makes them ideal for brushing excess flour off of bread dough during production or applying egg wash to dough before it is baked.

Rolling Pins

Rolling pins are used to roll out doughs. They come in various shapes, sizes, and materials. Rod and bearing rolling pins have a cylinder made of hardwood with a steel rod inserted through the middle, which is fixed with ball bearings and handles at either end. These heavy pins are ideal for large quantities or stiff doughs. They can be made of wood, stainless steel, or marble, though wood is ideal for bread production because it is not cool to the touch. Straight, or French, rolling pins are straight, thick dowels. They are traditionally hardwood, but are also available in nylon or aluminum. Their lack of handles makes it easier for the baker to feel the thickness and evenness of the dough during rolling.

Proofing Boards

These large wooden boards are thin and lightweight. They serve as a platform for shaped loaves during resting and proofing. Loaves can be placed directly onto the boards, or the boards can be lined with a couche (see below).

Couches

A couche is a heavy linen cloth used to protect loaves during proofing. The couche is placed on top of the proofing board, its excess hanging off one side. The shaped loaves are arranged inside the folds of the fabric and left to proof. The folds and the type of fabric help to preserve the integrity of the dough and keep it moist during proofing.

Proofing Baskets

Clockwise from top left: courrone banneton, round banneton, short oval banneton, long oblong banneton, lined banneton

Proofing baskets, or bannetons, have been used for hundreds of years and can also be used in place of couches to help preserve the shape of loaves during proofing. The baskets are made of plastic, straw, or willow, and are often lined with linen. Generally, they are available in round or oblong shapes. The baskets can also be placed atop proofing boards to make them easier to move around and store in large quantities during proofing. Willow proofing baskets are ideal because they allow excellent circulation and allow the dough to breathe. Some states do not allow bakeries to use willow baskets for food safety reasons, so be sure to check your local laws before purchasing them.

Covering the Dough

Bread dough should be covered through many of the steps of production. Direct exposure to air can cause some doughs to form a skin on the surface, and can expose the dough to elements of contamination (such as stray ingredients from another formula). Plastic coverings are ideal, particularly a thicker, heavier covering that protects but does not stick to the dough. Plastic wrap is suitable for products that are being stored in the refrigerator or freezer. Make sure the dough is fully covered according to the formula instructions.

Knives

A basic set of knives is useful in the bakeshop—for example, to prepare ingredients for the dough (such as inclusions and garnishes) or to slice finished loaves. Cutting utensils are available in different shapes and sizes, and are made from various materials. A chef's, or French, knife is an all-purpose knife. The blade is usually 8 to 14 in/20 to 36 cm long with a straight edge. Paring knives are short knives used primarily for peeling and trimming vegetables and fruits. The blades are 2 to 4 in/5 to 10 cm long, and they come in different shapes for different tasks. Slicers are available in several varieties. Some have offset handles, while others are straight; some have a straight blade edge, while others (especially for slicing bread) have serrated blades. Most are 8 to 12 in/20 to 30 cm long.

LAMES

Lames are thin arc-shaped razor blades clamped into a small handle that can be made of stainless steel, wood, or plastic. They are used to score proofed breads and rolls before they are baked (see page 107). The blade must be very sharp and be used in a swift, angled motion so as to create clean slices without pulling or tearing the dough.

Baking Pans and Molds

A variety of molds can be used in bread baking. Some are made of metal or flexible silicone and can be reused. Some molds are made of paper or aluminum and are disposable; when these are used, the breads are baked and sold inside of them. Common mold shapes include loaf pans, which are oblong or rectangular and available in a large range of sizes. Pullman loaf pans have sliding covers that fit on top, which produces a perfectly square loaf, ideal for slicing. Strap pans join four to five loaf pans together,

From the top left: muffin tins, strap pans, paper panettone pan, crimped paper mold, Pullman pan, crimped round bread pans, cornbread pan, à tête molds.

allowing multiple loaves to easily bake at the same time. Specialty molds are often used, if only for one type of bread. Tête molds are metal molds with fluted edges, used to create individual brioche rolls of the same name. Other common bread molds include panettone molds, stollen molds, and babka molds.

Sheet Trays

Sheet trays, or sheet pans, are flat rimmed baking pans ranging in size from 17¾ by 25¾ in/45 by 66 cm for a full sheet tray; to 12¾ by 17¾ in/33 by 45 cm for a half sheet tray; and 9½ by 13 in/24 by 33 cm for a quarter sheet tray. Their sides are generally 1 in/2½ cm tall. Sheet trays can be made of lightweight metal (with or without nonstick coating) or plastic. The metal varieties can be used for baking, but also for storage in the bakeshop (the primary use of the plastic trays).

Spray Bottles

When filled with water, spray bottles are ideal for spritzing the surface of bread doughs during proofing (to help keep them moist), before adding a garnish (to help it stick to the dough), or before baking (to help create steam in the oven).

Peels

There are two kinds of peels: the transfer peel and the oven peel. The transfer peel is a long, thin wooden board used to transfer loaves from their proofing storage to the oven loader or peel. The loaves are flipped (either from the proofing basket or by using the folds of the couche to move the loaf) onto the transfer board, and then the loaf is gently flipped again when placed on the loader or peel. Oven peels are made of wood or metal, and are used to transfer proofed loaves into the oven or to remove baked loaves from the oven. Generally, peels should be dusted with flour before the dough is placed on them. The dough should not be left on a peel or loader for too long, as it may begin to stick to the surface. The baker should use a quick, jerking motion to slide the dough off the peel.

3 INGREDIENTS

Most breads have very few ingredients—flour, water, yeast, and salt. This is one of the truly amazing things about bread baking: The same ingredients can be used in different quantities and manipulations to create a wide array of results. Nevertheless, to truly understand bread, a baker must fully understand each ingredient and its function.

FLOUR

Simply put, flour provides the structure for bread. It is the primary ingredient in all bread recipes, and it is the basis for the formulation of bakers' percentage (see page 78). In addition to the overall structure, flour is responsible for the contribution of a variety of characteristics to the finished loaf, including volume, break, grain, the color of the crust and crumb, texture, and taste. Bread flour, one of the most common flours used in bread baking, is derived from hard wheat, which is a high-protein wheat (between 11 and 13%).

Flour creates structure primarily through the formation of gluten. There are two water-insoluble proteins in wheat flour: glutenin and gliadin. These proteins become hydrated when they are mixed with water and begin to form gluten, which is responsible for gas retention in yeast-leavened products. The glutenin provides elasticity to the dough, the gliadin provides extensibility.

The gelatinization of the starches present in wheat flour also plays an important role in building structure. Starch crystals rupture at a temperature range of 140 to 180°F/60 to 82°C. This increased the surface area of each crystal, which is immediately surrounded by the available water present in the dough. Thus, the dough is transformed from its elastic, raw state to the more rigid structure of a finished loaf.

Wheat Flour

Wheat is the primary source of milled flours, and wheat flour is the most commonly used flour in bread baking. Even when flours from other grains are used, they are generally used in combination with wheat flour, which has high protein levels and produces gluten during mixing.

TYPES OF WHEAT

There are over 30,000 varieties of wheat, owing in large part to breeding programs and genetic engineering. These efforts were designed to produce wheat plants that are resistant to disease and produce higher yields. Wheat grown for flour production is classified by three characteristics: the climate of the region where it is grown, the hardness of the kernel, and its color. The classifications are used to describe the wheat itself, as well as the type of flour milled from it.

Hard wheat is used to produce stronger flours most suited for bread baking, while soft wheat is used to produce flours ideal for products with more delicate structures, such as cakes and pastries. Winter wheat is planted in the fall and is harvested in the summer. The plants lie dormant during the winter and then grow in the spring. This type of wheat has less protein overall, but the protein is of very high quality. When this wheat is milled, it produces a flour that has a better fermentation tolerance and greater dough extensibility. Spring wheat is planted in the late spring or early summer, and it is harvested in late summer or early fall. The growing season is shorter, so the plants are often treated with nitrogen to encourage growth. This produces a wheat with a very high protein content, but the quality of the protein is much lower than that of winter wheat. These flours produce less gluten and have a lower fermentation process.

This map highlights the large wheat-producing regions in the United States and their primary wheat varieties.

TYPE OF WHEAT	LOCATION GROWN	FLOUR USE
Hard Red Winter	Southern Great Plains	Bread
Hard Red Spring	Northern Great Plains	Bread
Soft Red Winter	East of Missouri and Mississippi Rivers	Pastries, Cakes
Hard White	Kansas	Bread
Durum	North Dakota, Montana, Minnesota	Bread, Pasta
Soft White	Washington, Oregon, Idaho	Flat breads, Cakes, Pastries

THE WHEAT KERNEL

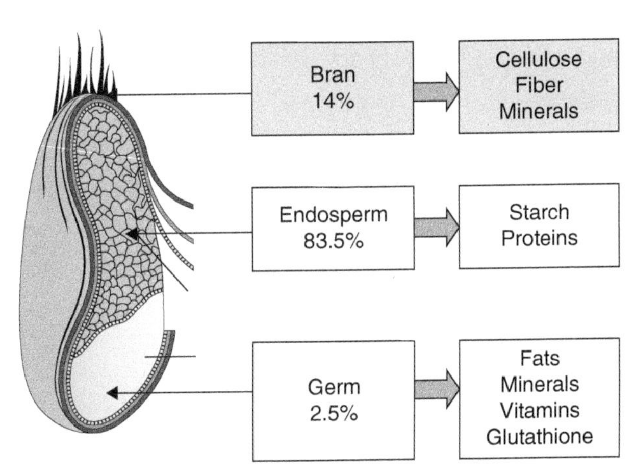

A kernel of wheat.

The wheat kernel, or wheat berry, is the seed of the wheat plant. The kernel is composed of three parts: the germ, the bran, and the endosperm. The germ makes up 2 to 3 percent of the kernel. This is the section from which the seed sprouts, and it is often separated from the kernel during the milling process because of its high fat content (around 10%). The higher the content of fat in a finished flour, the more prone it is to spoilage. The germ is high in B-complex vitamins and minerals. The bran makes up 13 to 15 percent of the kernel. The bran contains a small amount of protein, large quantities of B-vitamins, some minerals, and dietary fiber.

The endosperm makes up the bulk of the kernel (about 83%) and is the primary source of white flour. The endosperm is very high in protein, carbohydrates, and iron. It also contains some B-vitamins and some soluble dietary fiber. In the production of white flour, the bran and germ are separated during milling. These products are then milled and sold on their own as separate products. In the production of whole wheat flour, the entire kernel is milled to produce a darker flour, with all the nutritional value of each part of the kernel.

How Flour Is Milled

The process of milling wheat and other grains into flour is complex and time-consuming. During milling, various machines crack and separate the grains to produce the desired flour.

There are several ways flour is monitored during and after milling. The extraction rate is the percentage of flour obtained after milling. This varies with flour refinement (whole wheat flour, for example, has a 100 percent extraction rate because the entire grain is milled). The ash content is another milling standard that determines the mineral material remaining in the flour post-milling. Ash content is determined by burning a measured amount of milled flour and weighing the mineral material that remains.

While each flour is processed slightly differently, there are ten primary steps to milling any flour.

1. **Inspecting the grains.**

   Before the grains can be milled into flour, they are observed and tested. These tests are to determine the quality of the product and to check for any potential insect infestation. The results of the tests help determine what kind of flour should be produced from the grain. For example, different types of wheat are often blended to produce certain types of flours.

2. **Storage.**

   Grains are received in large quantities and often stored for a period of time before they are milled. On receipt, the grains are transferred to large storage bins, which are kept in rooms where the temperature and humidity are closely monitored to protect the grains from spoiling, mold, mildew, sprouting, and fermentation.

3. **Cleaning.**

   Before the wheat or other grains can be milled, they must be cleaned. A series of mechanized separators remove any foreign objects that might be mixed in with the grains. These separators use magnets to remove metal, screens to remove plant particles, an aspirator to vacuum away light impurities like dust, a disc separator to eliminate stones and other foreign particles based on size, and finally a scourer to clean and remove the outer husk of the grain.

4. **Tempering or conditioning.**

   The moisture content of grains upon arrival at the mill can vary, depending on the type of grain, the time of year it was harvested, and the region where it was grown. Before milling can begin, the grains must be tempered or conditioned to regulate the moisture content so that milling is consistent. Moisture is added to the kernels to strengthen the bran—this process eases the separation of the kernel's parts during milling. Tempering can last between 8 and 24 hours, with longer conditioning times needed to ensure slow, even water absorption.

5. **Grinding.**

   Grinding takes place in many steps. The grains first go through a piece of machinery called a break roll. The grains pass through a machine made of corrugated rolls that rotate toward each other. Each roll moves at a different speed to produce a shearing action and coarsely crack the grains. This separates the endosperm from the bran, which can be sifted out and ground separately. The grains then pass through additional grindings to obtain the desired texture.

6. **Sifting.**

   After the grinding, two steps are taken to ensure that the flour is fine enough. The first step, sifting, passes the ground grains through a series of varying sifters. Generally made of metal with either nylon or stainless steel screens, the sifter openings become smaller and smaller as the grains move from top to the bottom. This separates the larger pieces from the smaller ones; the larger ones are removed and ground again.

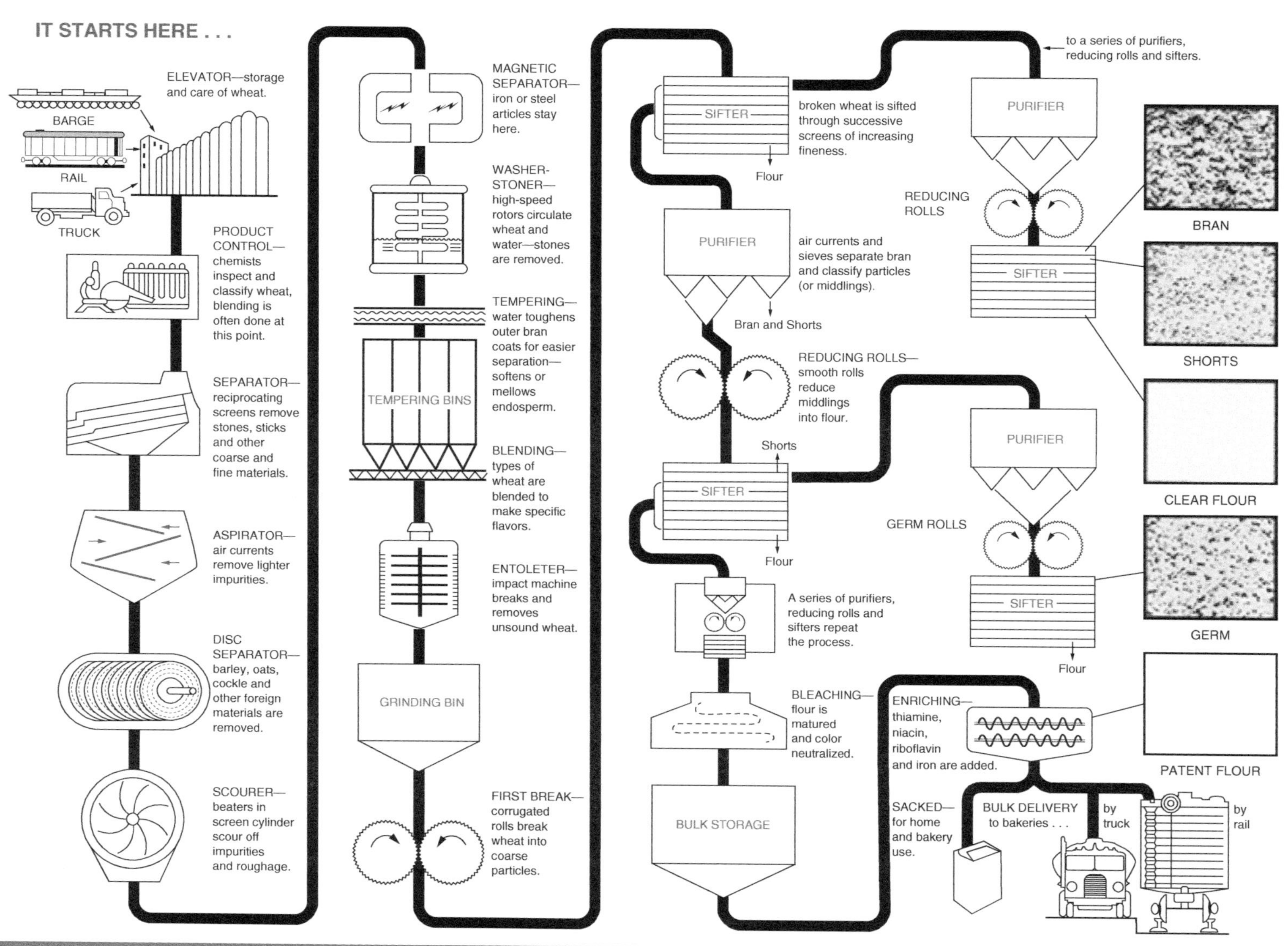

A simplified diagram of the milling process, from raw material to delivery.

7. **Purifying.**

   This step involves a passing of controlled air to ensure that the flour is properly ground. Any ground particles that contain portions of bran weigh less, are suspended, and then removed. They are then passed through a secondary grinding process to crush them again, then another series of sifters to remove as much of the bran as possible (unless whole wheat flour is being ground). The endosperm, once fully separated, is sent to the final, smoothest rolls for grinding into the finest texture, producing a fine flour. The germ is also successfully ground through these final rollers. Owing to its high fat content, it forms a large, flat piece. These pieces are removed in the final stage of purification. This entire process is repeated multiple times until the grain is properly ground into flour.

8. **Blending.**

   Not only are the various types of wheat different but crops can also vary vastly. Because of this, many mills choose to blend flours after they have been purified to achieve various results.

9. **Treating.**

   Some flours are treated to achieve a variety of results in the finished product. Newly milled, or green, flours cannot be used immediately after milling owing to the immature proteins found in the flour. Green flour absorbs more water and produces a less elastic dough. Freshly milled flours are also yellow, not white, in color. Flour should be aged for 2 to 3 months to whiten and properly develop the proteins. Some flours are bleached with chemicals that whiten the flour and replicate the aging process, allowing flour to be packaged and sold faster without an elongated aging process. Some flours are oxidized with one of several products to help with gluten development and improve the volume of the finished bread. Enzymes are added to some flours, usually one or more of the enzymes that are produced naturally by yeast (amylase and protease, see page 69), which are added to the flour to help produce an elastic dough that ferments faster after mixing. Often, enrichments are added to the flour to replace some of the nutrients that were lost during mixing—namely thiamine, niacin, riboflavin, iron, and/or calcium.

10. **Packaging.**

   Flour is generally packaged in bags of 5, 10, 25, 50, and 100 lb/2.26, 4.53, 11.33, 22.67, 45.35 kg. Finished flours are marked by milling date, batch number, and any treatments they may have received.

GRADES OF FLOUR

When flour is milled, it is given one of three designations: patent flour, clear flour, or straight grade flour.

PATENT FLOUR A flour made with a low extraction of grains, meaning most of the bran has been successfully removed during the milling process, is called patent flour. While this yields a very high quality flour, it is not ideal for bread baking because the bran produces a dough with greater extensibility and provides nutrients for the yeast to aid in fermentation. From this classification, there are two subgrades: extra short and long patent. Extra-short patent flour uses only the center of the wheat kernel, making it virtually free of bran (also classified as "low ash content"). Long patent flour uses more of the endosperm but also has been ground closer to the bran, increasing the overall ash content.

CLEAR FLOUR A flour made by continuing to grind the flour that remains after patent flour has been produced is called clear flour. This flour is made from the remnants that have been removed during the stages of milling patent flour, and therefore it contains more bran. This flour is most often blended by a miller to give more strength to rye or whole wheat flours.

STRAIGHT GRADE FLOUR A flour made through a standard, gradual grinding and sifting process is called straight grade flour. This flour can be packaged and used as is, or separated by the miller at the end of grinding to create both patent flour and clear flour. In general, straight grade flour is the most ideal for bread baking.

FLOUR SPECIFICATION SHEETS

Many mills provide flour specification sheets, or spec sheets, to those who purchase their products. These sheets provide detailed information about the batch of flour, which can help bakers achieve more precise results when preparing their formulas. Flour can be affected during every stage of the process, from the weather conditions during growing and harvesting of the grain, to each stage of milling, to the treatments added to the finished flour. The specification sheet explains how these factors led to the characteristics of the finished flour.

A **Mill and brand information**

B **Flour description:** Origin of wheat, type of wheat, certification of proper receiving, selecting, and cleaning of the product in compliance with laws.

Abbreviations for Flour Spec Sheets:

Hard Wheat: H	Soft Wheat: S	Durum Wheat: D	Winter: WW
Spring: S	Red: R	White: W	

C **Ingredient description:** Any additional ingredients that have been added to the flour. This can include one or more of the following:

- **Enrichments:** Vitamins, minerals, or other additions that alter the nutritional value of the flour.
- **Potassium bromate:** An oxidizer that is added to strengthen gluten development, fermentation tolerance, and volume of the finished dough.
- **Ascorbic acid:** Another oxidizer, but with the added negative effect of reducing extensibility in the dough.
- **Benzoyl peroxide:** Added to bleach the flour to whiten it, but can negatively affect the crumb color and the flavor of the finished bread.

- **Asodicarbonamide (ADA):** A bleaching agent that also acts as an oxidizer that speeds up the maturing process of flour before packaging.
- **Malt:** Contains an enzyme that aids in yeast growth and fermentation.
- **Fungal amylase:** Enzymes that aid in yeast growth and fermentation.

D **Test results:** A variety of tests are performed on finished flours to determine more information regarding their composition. These tests include:

- **Protein content analysis:** A measure of the total protein content of the flour, which is vital for bakers to determine if it's suitable for bread production. A combustion test is performed on a sample of the flour to determine its nitrogen content, which is translated as a percentage that can be applied to the entire batch of flour.
- **Moisture content analysis:** There can be no more than 14 percent moisture in a batch of finished flour just after milling. Packaging and storage conditions could alter the moisture content down the line.
- **Enzyme activity analysis (falling number):** Enzymes are naturally present in the germ and are responsible for breaking down the nutrients in the kernel to sprout and continue to produce life. The stage of germination at harvest determines the enzyme levels present in the flour (the enzyme activity travels to the endosperm when it germinates). The flour sample is mixed with water in a test tube and heated in boiling water. The mixture is stirred with a long, thin apparatus until it begins to coagulate. When it's fully coagulated, the stirring apparatus stands at the surface of the mixture, and the agitation stops. This process stimulates the enzyme activity and then causes them to begin breaking down. The amount of time (measured in seconds) it takes for the stirring apparatus to fall through the liquid is referred to as the falling number; it helps determine a flour's fermentation tolerance. An ideal falling number for bread baking is between 250 and 350.
- **Ash content analysis:** A test to determine the amount of bran present in flour just after milling. A sample of flour is incinerated to 1,652°F/900°C, which is a high enough temperature to burn off everything but the ash left behind by the minerals present in the sample (which are found in the bran).

E **Farinograph information:** The farinograph is a machine that measures the quality of a flour by providing the following information:

- **Dough development time:** Also known as "peak time," this is a number (measured in minutes) that expresses the amount of time it takes the flour to come to the proper consistency and development after water has been added to it. The higher the peak time, the more mixing is required to properly develop the dough.

Product Specification/ Technical Data Sheet

HARVEST KING - WHEAT FLOUR, ENRICHED, MALTED				
UPC	Code	Size	Mill Code	Revision Date 06/06/11
100 16000 53722 1	53722	50#	AV GF VN	HARVEST KING ENR MT ING Code 249896

DEFINITION

- This product shall be of food grade and in all respects, including labeling, in compliance with the Federal Food, Drug and Cosmetic Act of 1938 as amended and all applicable regulations there under. It shall meet FDA Food Standards for Enriched Wheat Flour as found in 21 CFR 137.165.
- A high quality patent bread flour milled from a selected blend of hard wheat. Wheat selection is to be consistent with optimum baking characteristics and performance. Wide variations in the type of wheat utilized for this flour are not permitted. The flour shall be produced under sanitary conditions in accordance with GMPs.

PACKAGING/SHELF LIFE/STORAGE CONDITIONS/PALLET CONFIGURATION

1. The package consists of 50 lb. multi-wall paper bags.
2. Stored according to GMPs at <80°F and 70% R.H., the shelf life is 1 year from the date of manufacture.
3. To preserve quality, dry storage at room temperature with regular inspection and rotation is recommended.

Size	Bags/Pallet	Bags/Layer	Gross Wt./Bag	Cube	Pallet Dimension
50# (West)	55	5	50.5	1.3	53"H/41.5"W/52"D
50# (East)	50	5	50.5	1.3	48"H/41.5"W/52"D

PHYSICAL CHARACTERISTICS

1. Color – Clean, creamy white, free of excessive bran specks.
2. The product shall be free of rancid, bitter, musty or other undesirable flavors or odors.
3. The product shall be as free of all types of foreign material as can be achieved through GMPs.
4. Falling Number – 240 – 280 sec.

KOSHER APPROVAL: Orthodox Union

ALLERGEN INFORMATION: Allergen - Wheat

INGREDIENT LEGEND:

Wheat flour, malted barley flour, niacin, iron, thiamin mononitrate, riboflavin, folic acid.

CHEMICAL COMPOSITION (14.0% Moisture basis)

1. Moisture	14.0%	Maximum
2. Protein	12.0%	+/- 0.3%
3. Ash	0.52%	+/- 0.03%

TREATMENT:

1. Enriched
2. Barley Malt

NUTRITION (Approx. per 100G)

Calories	357		Thiamin (B1)	0.64	mg
Protein	11.9	g	Riboflavin (B2)	0.40	mg
Fat	1.0	g	Niacin	5.30	mg
Saturated	0.14	g	Folic Acid	0.15	mg
Trans Fat	0.0	g	Iron	4.40	mg
MonoUnsaturated	0.08	g	Sodium	1.0	mg
PolyUnsaturated	0.45	g	Potassium	105	mg
Carbohydrate	73.1	g	Phosphorus	95	mg
Complex	71.7	g			
Sugars	1.4	g			
Dietary Fiber	2.9	g			
Soluble	1.8	g			
Insoluble	1.2	g			

MICROBIAL GUIDELINES: Listed as guidelines as opposed to controllable specifications

Standard Plate Count	<50,000/g
Coliforms	<500/g
Yeast	<500/g
Mold	<500/g

Do not eat raw dough or batter.

In this product specification sheet, the manufacturer provides details that may affect the product and how it is used.

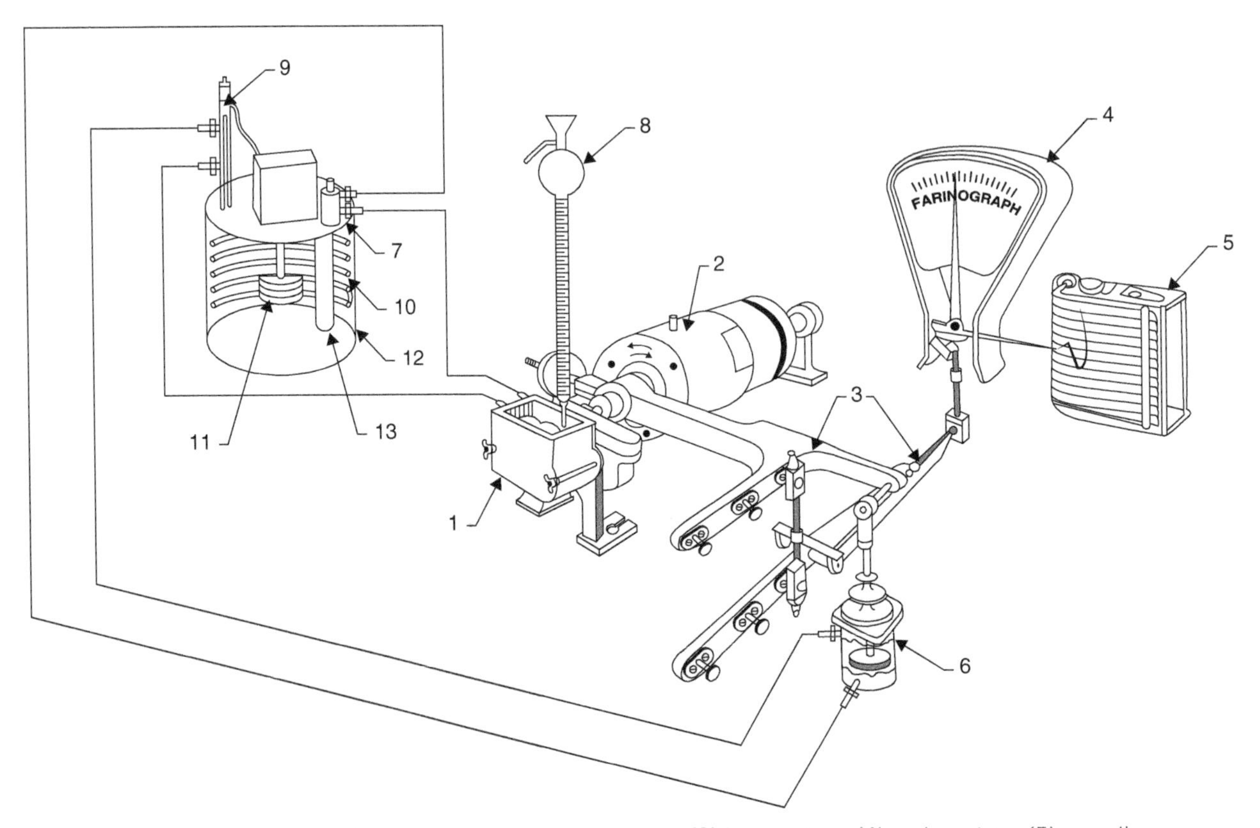

The basic parts of a farinograph: (1) mixing bowl, (2) dynamometer, (3) lever system, (4) scale system, (5) recording mechanism, (6) dashpot, (7) thermostat, (8) buret, (9) thermoregulator, (10) cooling coils, (11) circulating pump, (12) reservoir tank, (13) heating element.

- **Moisture absorption:** Presented as a percentage, this number reflects the flour's ability to absorb moisture during mixing. This is useful for bakers to reference when determining if formulas need to be altered slightly with new batches of flour.
- **Mixing tolerance index (MTI):** Represents how long a dough can be mixed before the gluten will break down.

F **Alveograph information:** Provides information on the dough's extensibility, elasticity, and overall strength.

- **Elasticity:** Expressed by the letter "P" and a number. A P value of 90 or above represents a strong flour with good elasticity; 50 or lower represents a weak flour with low elasticity.
- **Extensibility:** Expressed by the letter "L" and a numerical value. The lower the number, the less extensible the dough is.
- **Ratio of elasticity and extensibility (P/L):** A P/L ratio of around 0.70 is an ideal ratio of elasticity to extensibility, but it is often difficult to find the correct balance. This information can still provide bakers with information on how to handle breads during the later stages of production.

- **Strength:** Expressed by the letter "W" and a number. A W of 300 or greater indicates a strong flour; 200 or lower is a flour too weak for bread baking.

G **Additional information:** This can include classifications such as whether the flour is kosher or identify the flour as organic, gluten-free, and so on. Alternatively, it can offer storage suggestions or information about the mill not expressed elsewhere on the spec sheet.

Types of Flour

Most yeast-raised breads require a long fermentation time to develop the desired flavors. Not all flours can handle extended periods of fermentation without breaking down, though. If the structure of the dough breaks down, the bread won't rise properly in the oven and will have poor appearance and texture. While it is important for flours used in bread baking to have a high protein content, it is more important that the quality of the protein be strong enough to handle extended fermentation.

WHEAT FLOUR

WHOLE WHEAT FLOUR Milled from winter or spring wheat, utilizing the entire wheat kernel (including the bran and the germ), this flour has a protein content of 14 to 16 percent. Because of the high quantity of its protein, this flour will absorb more water, and formulas must be adjusted to account for increased hydration. It has excellent fermentation tolerance because the nutrients in the bran and the germ feed the yeast. It is available in fine, medium, and coarse varieties. It's very high in nutrients, but often produces the best results when combined with white flours. Owing to its higher fat content (in the presence of the germ), this flour is more prone to spoilage and rancidity.

HIGH-GLUTEN FLOUR Milled from spring wheat utilizing the entire endosperm section of the kernel, this flour has a protein content of 13 to 14 percent. While its protein content is high, its fermentation tolerance is low. It's most often used for breads in which a very thick, very chewy crust is desired, such as bagels or hard rolls.

BREAD FLOUR Milled from hard winter or spring wheat, this flour has a protein content of 11 to 12 percent. The flour has a high fermentation tolerance, making it ideal for most bread formulas, even if mixed with other flours for additional flavor or texture.

SEMOLINA AND DURUM FLOUR Milled from the durum wheat kernel, this flour has a protein content of 12 to 14 percent, and its color is golden yellow. Generally, this flour produces better results when combined with a white wheat flour, such as bread flour. It is sold in two varieties: semolina, which is more coarsely ground; and durum, which is finely ground.

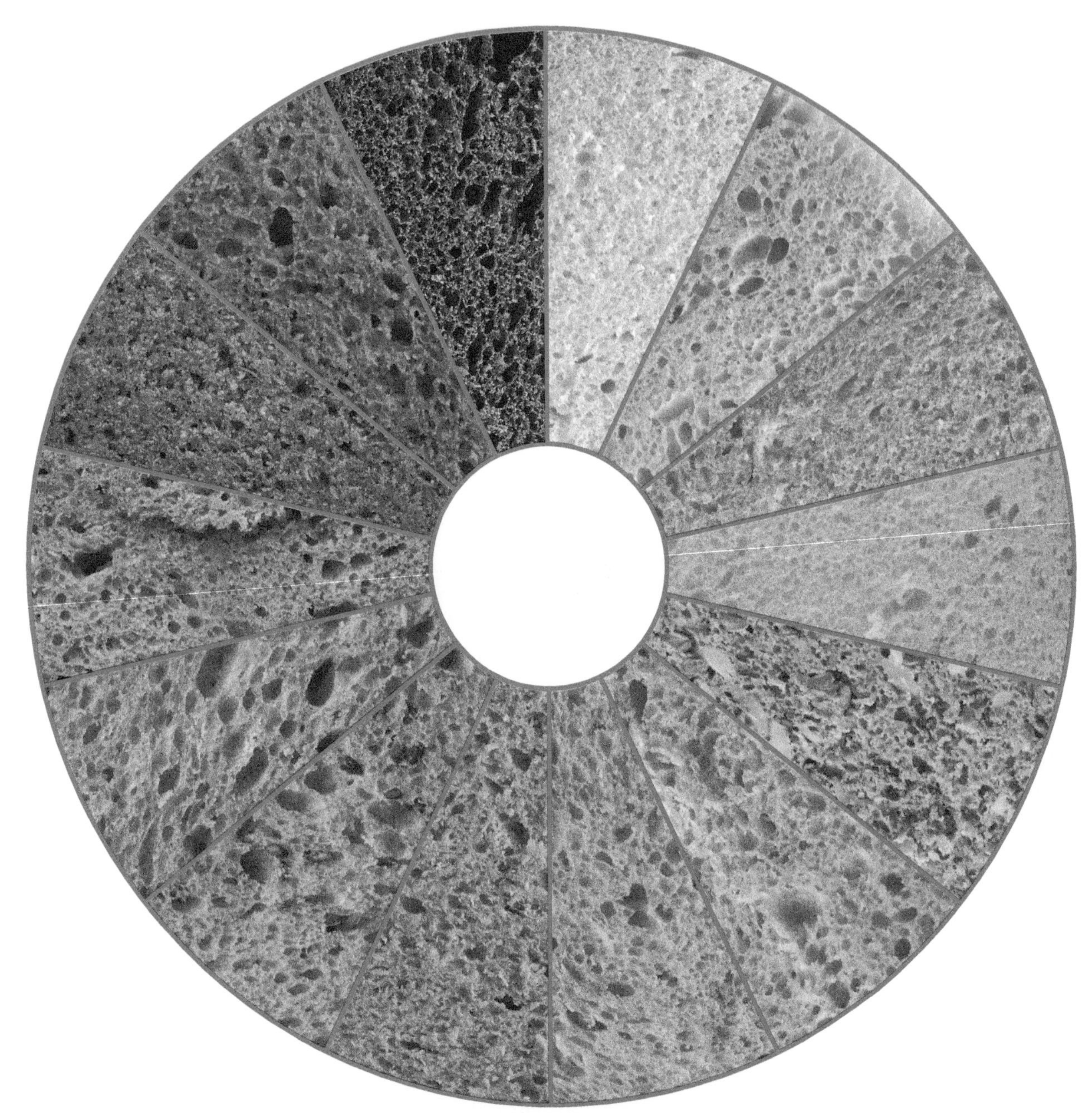

While people often think that the interior of bread is just one color, different flours can create a multitude of crumb colors and textures.

ALL-PURPOSE FLOUR Milled from a blend of hard and soft wheat flours (exact blends can vary by producers), this flour has a content of 8 to 12 percent. This mid-range flour has decent fermentation tolerance, but not as much as its higher protein counterparts. Not generally used in bread baking unless it is being combined with other flours in a formula.

PASTRY FLOUR Milled from soft winter wheat, this flour has a protein content of 8 to 10 percent. Its low protein content produces less gluten, making it less than ideal for bread baking and better for cookies, biscuits, and other pastries.

CAKE FLOUR Milled from soft winter wheat, this flour has a protein content of 6 to 9 percent. Generally bleached to produce a very white product, this flour is used specifically for cakes because its lack of protein makes it difficult to generate gluten, producing a more tender end product.

OTHER FLOURS, GRAINS, AND CEREALS

While wheat flour is one of the most predominantly used flours in the bread bakeshop, other flours can add flavors and textures. Rye flour, in particular, can be utilized to produce a wide array of traditional regional breads, but it is widely misunderstood because it behaves so differently from wheat flour. In addition to flour, other milled grains can be added to bread doughs or as a garnish on the exterior of a loaf or roll.

RYE FLOUR Milled from rye kernels utilizing a similar method as for wheat flour, rye is sold in several varieties based on color, strength, and protein content. Available in white rye, medium rye, dark rye, and pumpernickel, rye flours have a distinct flavor, color(s), and texture. Some of rye's unique properties require special understanding and application. For more information on rye, see Chapter 8.

BUCKWHEAT FLOUR Milled from either roasted or unroasted buckwheat groats, this grain is not wheat at all, and can actually be classified as gluten free. It has a strong, nutty flavor, which is even more pronounced when the grain has been roasted. The grain's strong flavor can be attributed to its high fat content, which makes it dark in color and resemble a whole-grain flour.

RICE FLOUR A gluten-free flour milled from white, brown, or sweet rice. Its favor is mild, and the flour is soft, with a low protein content, yielding tender baked goods.

TEFF FLOUR Milled from a very small grain native to Ethiopia, which has a nutty favor. Available in ivory, red, and brown varieties, it is very high in protein.

OATS Before packaging, oats are cleaned, toasted, and hulled. Cleaned whole oats are referred to as oat groats. Cracked oat groats are sold as steel cut (also known as Scottish or Irish) oats. Oat groats are steamed and flattened to produce the widely used rolled oats. Instant oats are made by precooking, drying, and rolling oat groats for a faster cooking process. Oat groats can also be ground finely to produce oat flour, which is gluten free.

CORNMEAL Milled from dried corn kernels, generally available in yellow or blue, depending on the color of the corn that is ground. Cornmeal is available in varying levels of coarseness—fine cornmeal is often mixed directly into doughs, while coarse ground cornmeal is more often introduced as a textural component or garnish to dough.

AMARANTH A small grain with a cornlike flavor. It is sold as a whole grain and has high starch and protein contents.

SPELT A grain containing more protein and less gluten than wheat. Available as a whole grain and also milled into flour.

QUINOA A small, quick-cooking grain with a lightly nutty flavor and unique texture. Available in white, red, and brown varieties.

FARRO An ancient grain related to wheat with a very high starch content.

MILLET A naturally gluten-free ancient grain that is available whole, cracked, or ground into flour.

BARLEY A common whole grain, very high in fiber, that is sold whole (most often as pearled barley) or ground into flour.

WATER

A seemingly basic ingredient, water plays crucial roles in bread baking. At the most simple level, water hydrates the flour, which is the basis of the dough. Flour is composed of both starch and protein, both of which are hydrated by the water, but at different times. When water is added to flour, it begins to form a dough when the starch is hydrated. As the dough is mixing, the protein in the flour absorbs the moisture, which begins the formation of gluten. Because the starch absorbs moisture more quickly than the protein, it is important to begin mixing the dough at a slow speed for proper starch hydration and dough formation. After the starch has absorbed the water, the mixer speed can be increased to encourage gluten development.

Beyond hydration, water kick-starts the enzymatic activity and the fermentation process in the dough. The amount of water in the formula also determines how these reactions develop—more water causes a dough to ferment more quickly, while less water causes the fermentation to develop at a slower rate. (Note: This permits formula adjustment in wet doughs to include less yeast so as to compensate for the faster rise.)

Water also controls the dough's temperature, which is vital to the fermentation process. Because it is easy to alter the temperature of the water in different types of weather or under different conditions, bakers use water to help achieve the desired dough temperature (see page 82). Cooler water might be used to compensate for warmer temperatures, very fast or extended mixing times, or to slow down fermentation. Warmer water might be used to kick-start yeast activity in cooler temperatures and speed up fermentation.

Water can also control the overall consistency of the dough. While it's one of many factors that can affect the final consistency (ingredient type, mixing time, and fermentation time are also included), the quantity of water in a formula can be altered to adjust the dough. Because flour conditions and characteristics can greatly alter the outcome of the finished dough, it's difficult to predict how much water is needed to achieve doughs of different textures. There is no precise percentage of water, for example, to achieve a stiff dough. Skilled bread bakers can identify the consistency by touch, however.

In general, a larger amount of water will create a well-hydrated dough with a softer consistency and less gluten structure. This dough will be more extensible, but less elastic, and so it should have a long fermentation time. Owing to the high level of moisture in the dough, the baked loaf will have an open crumb and chewy texture. The longer fermentation time also generally means that these breads have stronger flavor. Less water will create a stiffer dough—less hydration means a stronger structure of gluten will form, and the dough will be more elastic, but less extensible. Because the dough is stronger, it should ferment for less time, lest it become too strong during shaping. The baked loaf will have a firmer texture with a tighter cell structure, and the flavor will be more mild.

Calculating the Amount of Water for a Formula

When developing or altering a formula, there is a basic method for determining how much water is needed to properly hydrate the flour. It is generally recommended to start with a known quantity of water (between 60 and 65%). Begin mixing with this amount of water, and mix for 2 minutes. At this time, assess the dough: If the dough is too stiff, 1 percent additional water (meaning 1% of the original flour weight) can be added until the proper texture is achieved. The final quantity should be calculated by adding the initial water and the added water to create the standard for the formula. While this is a good starting point, bakers must remember that other factors can affect a dough's hydration and how the water is absorbed into the dough.

Water plays an important part in the mixing of the dough, but its role does not stop there. After bulk fermentation, the dough is divided, preshaped, and then shaped (see more on these steps in Chapter 4). Highly hydrated doughs are stickier and more difficult to shape. Ciabatta, for example, is handled very little, cutting the dough and stretching it gently to form a rustic loaf without excess handling. Stiffer doughs can be shaped more easily, but they are more prone to drying out. To prevent this, the dough should be properly covered during fermentation, especially if the air is overly dry. If a dough becomes sticky owing to excess moisture on the surface (rather than the hydration level or mixing style), this could be a sign of high levels of moisture in the air. If so, the dough should be left uncovered and given air circulation to prevent excess moisture.

Water also plays an important role in the baking process. First, the hydrated starch molecules become gelatinized at a certain temperature, turning the raw dough into a baked loaf. The water molecules inside the dough travel from the outside of the starch particles (where they have hydrated them, bringing the dough together) to the inside as the dough bakes. This creates the interior structure, or crumb, of the loaf. During this same time, the strands of gluten, also providing structure to the loaf, dry out and set into their final composition. The formation of the crust also occurs as the exterior of the dough fully dehydrates, creating a crisp and/or chewy exterior.

One of the most important things that water provides during baking is adding steam to an oven. Steam can be introduced in a variety of ways during baking, but its role can be pivotal. The steam condenses on the surface of the dough, and helps to produce a crisper crust, while also promoting better browning of the crust (which provides texture, visual appeal, and better flavor).

Water continues to play a part even after baking. During the cooling process, the baked loaf continues to release moisture. If the bread is properly cooled, the excess moisture is absorbed by the surrounding air. If the bread is not allowed proper ventilation and air circulation during cooling, the steam produced by the finished loaf will be reabsorbed into the crust, making it soggy. Moisture loss continues as the bread becomes stale as well. Proper cooling and storage keep the bread at its highest possible quality for the longest period of time.

Water as an Ingredient

Beyond its many functions in the baking process, water is an ingredient, and it contributes to the bread as such. Only good-quality, drinkable water should be used in the bakeshop, but it is important to remember that modern drinking water has often been treated for a variety of reasons, and these treatments can affect the dough in several ways. Filtering the water can help minimize negative effects.

FLAVOR

Any detectable flavors or smells in the water can end up in the finished bread. In general, negative tastes and aromas can be removed through proper filtration. Some of the minerals and chemicals in the water can also create strange flavors in the bread.

MINERALS

Minerals are added to modify the hardness and softness of the water. The primary minerals added are calcium, sodium, and magnesium. Yeast absorbs and feeds on these minerals as nutrients, and as such, their presence in the water can strongly affect fermentation. Hard water is water with higher than average levels of acids, as well as iron and other metals. These acids and minerals can speed fermentation by providing larger quantities of nutrients to yeast. In addition, higher sodium levels in hard water can bind the sodium to the gluten structure and overly strengthen the dough. A water softener can be used to control the strength and mineral content of hard water, but it is important that some minerals remain to supply the yeast with nutrients. Another option is to decrease the amount of supplemental food for the yeast in the formula (such as sugar or enrichments). Soft water has a lower mineral content, leading to slower fermentation and a weaker structure. To combat this lower mineral content, the formula can be altered (more yeast food, additional salt, or the addition of dough improvers) to compensate for the slower fermentation time.

Note: Some bakeries install water softeners in the water supply lines that feed their proofers and/or oven steam generators. This can help prolong the life of the equipment and reduce overall maintenance by preventing the accumulation of mineral deposits.

SOLIDS AND TURBIDITY

Water can contain solids that can alter yeast growth and fermentation in the same way as does hard water. These elements can vary and are sometimes only present in trace quantities, needing precise testing and analysis to detect. Turbidity is the occurrence of suspended solids in the water, which makes it look less clear. The higher the turbidity, the more likely it can affect the dough, resulting in accelerated fermentation time and alteration of final crust and crumb color.

CHEMICALS

Chemicals are often added to water to make it safe to drink. Most notably, chlorine is added to much of America's drinking water. Yeast is sensitive to chlorine, so high levels of chlorine can inhibit or prevent its growth. Contact your local water board to check on the level of chlorine in the water. And because enzymes are added to many varieties of flour, chlorine's affect can stretch to the flour as well. Water should be tested to ensure that its chlorine content is less than 10 parts per million (10 PPM). Any more than that and the chlorine will greatly alter fermentation.

Water Quality and Reverse Osmosis Systems

If bakers have concerns about water quality, they may consider investing in a reverse osmosis system, which is an advanced filtration system for water, reducing its chemical content and impurities. The main advantage of such a system for a bakery is that it keeps the water completely consistent, year-round. However, while consistency is important to most bakers, water quality is less of a concern in the United States, making expensive equipment like this not necessary in terms of safety. Bakers can learn to alter their formulas with changes in the season and weather that might affect the water supply. In addition, with fewer concerns about water safety, bakers can assume that problems with a bread formula more likely stem from scaling, mixing, or handling.

pH

Yeast cannot thrive in an alkaline environment, meaning that water with high pH levels can drastically slow or halt fermentation. However, water with high pH levels can be altered by using vinegar or lactic acid. Commercial dough improvers (see page 59) can also be added. Water with especially low pH levels can overly accelerate fermentation, which may require reducing overall fermentation and/or altering the formula to increase the food source for the yeast. For extreme hard-water conditions, add 0.1 percent ascorbic acid (vitamin C) to the water.

YEAST

Yeast is a living organism responsible for much of the flavor and texture of bread. Yeast was discovered during the study of winemaking and was found to be the cause of the fermentation of the grapes. Further studies helped scientists isolate and refine pure cultures of yeast so that they could be added directly to a recipe to produce fermentation. The strain of yeast sold as commercial baker's yeast is called *Saccharomyces cerevisiae*, and it has been determined to be the best suited for bread baking, though it is also used in the production of beer and other alcoholic beverages.

Yeast needs suitable conditions to thrive. First, it is important to note that yeast can survive in two environments: aerobic (with air) and anaerobic (without air). In an anaerobic environment, the yeast is held in suspension with no activity; this is why yeast is packaged today in airtight, vacuum-sealed bags. In an aerobic environment, the yeast begins to feed on available carbohydrates and multiplies. This multiplication is important in commercial yeast production (see page 62). To thrive, yeast needs warmth, moisture, and food to begin fermenting. In the proper environment, the yeast will flourish

The Basics of Fermentation

Fermentation is vital to building the internal structure of bread dough. Fermentation is defined as the anaerobic respiration of microorganisms. The process converts carbohydrates into alcohol, carbon dioxide, and a variety of organic acids. The carbon dioxide leavens the bread by trapping gas in the protein-rich glutinous strands developed in the dough. The alcohol tenderizes the gluten strands to improve the overall texture of the bread, and then is cooked out during baking. The organic acids (including glycerol and aldehyde) contribute to the flavor and aroma of the finished loaf.

Additionally, fermentation provides flavor to the finished loaf. Doughs that ferment slower for longer periods of time will have a more pronounced, stronger flavor than doughs that ferment quickly. This is especially evident through the use of preferments, which utilize naturally occurring wild yeasts (see page 121).

If not monitored and/or controlled by the baker, fermentation can continue until the yeast runs out of food and stops producing alcohol and carbon dioxide, or until these by-products kill the yeast. Bakers should understand the three components that factor into fermentation:

- **Temperature:** Yeast thrives in a temperature between 80 and 90°F/27 and 32°C. Temperatures below 80°F/27°C will retard, or slow, the yeast's development. This can be desirable if doughs are being mixed, fermented, and shaped ahead of time for baking at a later time. Temperatures above 105°F/41°C will also slow fermentation and temperatures 138°F/58°C or higher will kill the yeast completely.
- **Moisture:** Yeast cannot survive in a dry environment. Yeast becomes hydrated during the mixing of the dough, and will continue to thrive unless the dough becomes too dry. Bakers should monitor the bakeshop environment and keep the bread dough covered, especially if the air is dry.
- **Food:** Yeast feeds on carbohydrates, which are naturally present in the bread dough. Many recipes call for additional ingredients such as sugar to help kick-start yeast growth and thereby quickly beginning fermentation. But too much additional food, sugar especially, can actually slow the fermentation process. Sugar is hygroscopic and can draw moisture out of the environment (salt has a similar effect; see page 65), so it should not come into direct contact with the yeast or be used in too large quantities, so as to avoid any negative effects.

and ferment properly. Too much or little of any of these elements, and the yeast will be negatively affected, resulting in poor fermentation.

In bread baking, the yeast is subjected to both aerobic and anaerobic environments at various stages of the process. There is a small amount of oxygen present in the dough at the time of mixing (air is incorporated throughout this process) until about 15 minutes into bulk fermentation. This provides an opportunity for the yeast to begin feeding and begin reproducing. While the yeast may start to multiply during this time, it is more likely that the oxygen will be gone before the yeast has had time to bud. In an anaerobic environment, the yeast cannot simply feed on the sugars present in the dough and easily reproduce. The yeast has less energy without the presence of oxygen, and is forced to convert available food into alcohol and carbohydrates, thus beginning the process of fermentation. Studies have revealed that there are over 200 identifiable aromas present in bread dough as a result of a lengthy fermentation period. While not all of these aromas survive the baking process, it does explain why modern artisan breads have such complex flavors and aromas.

Dough Maturation

In addition to leavening and producing flavor, fermentation allows the dough to mature, an important component for developing the internal structure of the dough. Maturing the dough strengthens the gluten bonds by exposing them to the organic acids that are a by-product of yeast development. Additionally, the gluten strands are stretched and extended owing to the presence of carbon dioxide. During folding (see page 99), the dough is manually stretched as well, redistributing the yeast food as well as the carbon dioxide, further extending the gluten strands.

Note: Doughs with a shorter mixing time require lengthier fermentation/maturation periods than doughs which are more developed because of longer mixing times.

Yeast Production

Yeast is produced commercially by providing the most ideal environment for the yeast to reproduce. The multiplication of yeast is called "budding." The "mother" yeast cell begins to grow a second yeast cell on one side that eventually splits off. This offspring can also multiply in a similar fashion. In 5 days, the yeast can reproduce to over 100 times its original quantity.

In its early stages, yeast is grown in a mixture of molasses and water. When the mixture arrives at the factory for large-scale production, the molasses is removed through a sterilization process and the molasses is cleaned to certify it as food grade. After cleaning, it is combined with water to form the wort. The manufacturer then takes a strain of yeast (strains vary from producer to producer), and combines it with the wort to begin producing yeast. This is done in a very small quantity; as the yeast multiplies, the mixture is transferred to larger and larger containers called "fermenters." The entire process is closely monitored and levels of various elements (sugar, nutrients, oxygen, temperature, pH) are diligently regulated to ensure that the finished product has the ideal characteristics. The speed of growth is primarily regulated by the amount of food the yeast has access to. Additionally, air is constantly pumped into the fermenters to provide ideal levels of oxygen for reproduction.

Eventually, the yeast reaches the proper level of concentration, potency, and quality to be used for commercial baking. At this point, the yeast is separated from the wort and sent to be packaged.

Types of Yeast

While there are many strains of yeast, there are six primary varieties of yeast that are sold and used in baking. They all work, but they are used differently.

FRESH YEAST

Fresh yeasts are yeasts packed and sold with minimal processing, meaning they are not dried for packaging. These yeasts have a shorter shelf life than dried yeasts, and their moisture levels make them somewhat less predictable to use.

CREAM YEAST

The first and least processed by-product of yeast production, cream yeast contains only about 20 percent solids and 80 percent moisture. This yeast is highly perishable and must be used very shortly upon production. Therefore, this yeast is generally used only by bakeries located fewer than 1,000 miles from the yeast factory. The yeast is delivered refrigerated and must be stored under refrigeration until the moment it is added to the mixer. Its shelf life is only 2 weeks, but because it is in liquid form, it disperses quickly into the dough, making it easy to mix. Here, 1 gallon/3.74 L of cream yeast is equivalent to 5 pounds/2.27 kg of compressed fresh yeast.

COMPRESSED YEAST

Compressed yeast is the result of the next stage of processing, where the yeast is processed through a filter to remove some of the moisture. The result is sheets with around 30 percent solid content and 70 percent moisture content. The sheets are either crumbled or cut into blocks and packaged. Compressed yeast should be refrigerated at all times. Its shelf life is slightly longer—3 to 4 weeks from the date of processing. The texture should be moist and firm, with a light brown color. A dry appearance or dark color means the yeast is spoiling and/or has been stored improperly.

DRIED YEAST

Dried yeasts have a long shelf life and are therefore easier for bakers to store. Generally speaking, dried yeasts are preferred by most bakers owing to their ease of use and consistency.

ACTIVE DRY YEAST Active dry yeast is dried to contain a moisture content of 7 to 9 percent. It may be treated before drying to aid in the process. First, the yeast is extruded in thin strands, and then it is further dried to produce individual spheres of dried yeast. These spheres are actually active yeast cells surrounded by dead yeast cells. Therefore, the yeast must be rehydrated before use in warm water (110° to

115°F/43° to 46°C) for about 15 minutes to remove this outer layer and activate the yeast. If the yeast is rehydrated in water that is too warm (above 120°F/48°C) or too cold (below 40°F/4°C), the yeast will die. The yeast can survive to below 40°F/4°C, but its activity will be severely diminished, slowing fermentation drastically. The activated yeast can be added at any stage of the mixing process, but it is generally added early on to ensure proper incorporation into the dough. Active dry yeast has a very long shelf life and can be held for up to two years at room temperature, unopened. Once opened, the yeast will survive up to 3 months, refrigerated, or up to 6 months in the freezer. If the yeast is stored at room temperature after it is opened, it will lose its potency the longer it is stored.

INSTANT DRY YEAST A more recent addition to the market, instant dry yeast is manufactured similarly to active dry yeast, with a few exceptions. The yeast is dried much quicker than active dry yeast, with a finished moisture content of about 5 percent. This process produces less dead yeast than active dry, allowing the yeast to be mixed directly with the flour rather than needing to be rehydrated first. Because there is less dead yeast, bakers can use 25 percent less yeast in a formula than when using active dry, and 30 percent less than when using compressed yeast. The drying process make the yeast sensitive to moisture and oxygen, so it is packaged in vacuum sealed packaging to improve its shelf life. Unopened, it can keep at room temperature for up to 2 years. Once opened, it should be kept in a sealed container and can survive for up to 6 months. Instant dry yeast is more sensitive to water temperature and should never be mixed with water less than 70°F/21°C. Temperatures this low will cause the yeast to release large amounts of its glutathione, an enzyme that provides extensibility and promotes fermentation.

Note: Another type of instant yeast has been developed especially for use in doughs that will be frozen for over 2 weeks. This yeast is sold frozen and should remain in the freezer except during mixing.

DEACTIVATED YEAST

Deactivated yeast is a dead yeast with very high levels of glutathione. It makes an extensible dough that is very easy to shape and manipulate. But because the yeast is not alive, it does not feed on carbohydrates and produce fermentation. Therefore, this yeast is only added in formulas as an additional ingredient to increase the consistency of the dough.

SALT

Salt is widely known for its flavor-boosting abilities. But to the bread baker, salt is much more—playing important roles in each step of the baking process. This was not always the case, though. Throughout history, salt was often too expensive to be used by the average baker. Even when it became more economical, it was used sparsely, mostly to condition or treat the wheat in years that produced bad crops. Some of the earliest French cookbooks used only 0.45 to 1 percent. In America in the 1950s, salt use increased to 1.5 percent. But the introduction of convenience methods and products in the 1960s changed the way bread was made. More intensive mixing methods caused bakers to increase the amount of salt to compensate for a lack of flavor in a longer fermented dough—as high as 2.5 percent. Today, the average amount of salt in most artisan bread baking is around 2 percent.

During mixing, salt comes in contact with the proteins in the flour and creates bonds that produce a more stable dough and gluten structure, yielding a stronger dough overall. Salt is also hygroscopic, which means it plays an important role during mixing, drawing the water away from the surface of the dough and thus limiting tackiness or stickiness, thereby making the dough easier to handle and shape.

Because salt has a natural ability to slow down biological reactions, it is responsible for maintaining the proper fermentation speed. The salt slows and/or regulates the fermentation, allowing the baker to control the yeast activity to maintain the proper flavor development. Salt also promotes gas retention because the dough has developed increased strength during mixing. This increased strength is combined with good extensibility, meaning that, after fermentation, the dough is easier to properly shape.

Salt's hygroscopic properties also play a role in baking, providing humidity on the dough's surface, thereby keeping it moist during baking for a longer period. This ensures a thinner crust, which is therefore darker, shinier, and crisper. And after baking, salt can affect the shelf life of the finished loaf. In a dry environment, the salt keeps the bread moister longer and delays the staling process. In a humid environment, however, the salt's hygroscopic tendencies attract humidity from the surrounding air, and the bread becomes soggy and eventually moldy.

How and When to Add Salt

Air is incorporated into the dough during mixing. This introduction of oxygen reinforces the gluten bonds, but an excess of air can negatively affect some of the composition of the flour in the dough. Salt can impede this negative effect by slowing the reaction if it is added at the beginning of mixing. If salt is added later in the mixing process, though, it will not interfere with oxidization. The one exception to this rule is the process of autolyse (see page 93), during which salt is added after an initial stage of mixing. This is done intentionally to ensure that the salt will not interfere with the enzyme activity in the dough.

Note: Salt's hygroscopic nature and ability to slow down reactions can have a negative effect on yeast, and therefore should never come in direct contact with yeast. It is best to mix the yeast into the flour before adding the salt.

Types of Salt

Different types of salts have varying textures and levels of salinity. Most of the formulas in this book call for kosher salt, but other salts might be used as finishing salts on the surface of the bread.

TABLE SALT

This salt may contain iodine. Its small, dense grains adhere poorly to food and do not dissolve easily.

KOSHER SALT

Kosher salt is a coarse salt that weighs less by volume than table salt. It dissolves easier and is free of any additives, giving it a cleaner taste.

SEA SALT

Sea salt is collected through the evaporation of seawater. It is available in varying levels of coarseness. Fine sea salt is often used in bread baking. The coarser grains are more often used as finishing salts. While this salt is derived from seawater, it is processed in an industrial plant, which means it can still contain bleaching agents and chemicals like iodine.

CELTIC GRAY SALT

This is a true sea salt from Brittany, in northern France. It is raked from mineral-rich basins, harvested by hand, and left completely unrefined. Its gray color comes from its high mineral content. These minerals boost the flavor and also allow the human body to absorb the salt more easily than refined salts.

FLEUR DE SEL

This is the finest of the sea salts harvested where the Celtic gray salt is found. That is, the first salt skimmed from the water's surface is sold as fleur de sel. Its color is whiter because it has had less contact with the minerals in the floor of the basin. Its texture is also coarser and flakier.

HAWAIIAN PINK SALT

A coarse salt harvested off the coasts of Hawaii. The basins where it is harvested are rich with a clay that is high in iron oxide, which gives the salt its signature pink or dusty red color.

BLACK LAVA SALT

Also originating in Hawaii, black lava salt is harvested from ocean waters and treated with coconut-shell charcoal to augment its black color and intense flavor.

SEL DE GUÉRANDE

Harvested in northwestern France, sel de Guérande is a coarse natural sea salt harvested in the salt marshes of the Guérande Peninsula. Water is naturally evaporated to produce this flaky sea salt.

SMOKED SALT

This is a medium-grain to coarse salt that is smoked over alderwood or oak from French wine barrels. The finished salt is strongly flavored and has a light brown or gray tint.

OTHER INGREDIENTS

While many breads can be made with just the four simple ingredients listed above, many breads have additional ingredients that provide additional flavors, colors, textures, and other effects on the finished bread.

Enrichments

These ingredients are added to doughs to contribute to their color, texture, and flavor. While many doughs that contain just flour, water, salt, and yeast are called lean doughs, breads made with these ingredients as well are called enriched doughs.

EGGS

Eggs improve the condition of doughs. They also give color and affect browning. When they are used in dough, eggs always include the yolks; whites alone make the crumb dry and accelerate the aging process. Doughs in which eggs are used include brioche, challah, and yeast doughnuts.

SUGAR

Sugar is added to doughs to feed yeast and aid in fermentation. In addition to feeding the yeast, however, sugar contributes to the appearance of the final loaf as a result of caramelization, especially in the crust. Residual sugars that remain after fermentation also contribute to the flavor of the bread, determining if it is sweet or not.

Many varieties of sugar can be used in bread baking, including granulated varieties of cane sugar, sucrose, and dextrose, or liquid syrups like malt syrup, corn syrup, or honey.

Types of Sugar:

- *Granulated sugar*: white sugar with a crystal size ranging from coarse to medium to fine
- *Liquid sugar*: solution of highly refined sucrose or invert sugar
- *Glucose syrup*: made from cornstarch
- *Sugar substitutes*: sorbitol, xylose, dietary baking products

BUTTER

Butter contributes to the flavor, color, and mouthfeel of the finished dough by enriching it with fat and milk solids. The fat content also slows the staling process by keeping the finished bread very moist, making the finished loaf very easy to slice. Shortening, lard, and oils may also be used to achieve similar effects, though they contribute different flavors to the finished loaf.

MILK

Milk contributes protein, carbohydrates, and minerals to a bread dough. In turn, these play a role in many parts of the finished loaf. The lactose feeds the yeast, improving fermentation; the minerals enrich the bread, improving extensibility; and the protein improves overall dough strength and contributes to the finished appearance of the loaf by promoting browning. In addition to regular milk, blends of soy and whey are often used because of their higher protein and lactose contents.

WHEAT GLUTEN

Wheat gluten can be processed into a flour-like powder and sold separately. Some bakers add it to weaker flours to improve their overall strength by increasing the protein content and absorption.

MALT

Malt is made by separating the enzymes that break down starch into sugar from cereal grains, usually barley. Many flours are treated with malts at the mill, but some bakers add malt to formulas because of its impact on the dough. It is available in two varieties: diastatic and nondiastatic malt. Diastatic malt is sold in either syrup or dry flour form. Diastatic malt provides additional food for the yeast and improves dough handling. Nondiastatic malt is heat treated and is sold as a syrup. It contributes to the browning of the crust, contributing flavor and color and also promoting fermentation.

Commercial Enrichments

Large commercial bakeries often add enrichments to their doughs to produce specific reactions and contribute to the desired end result.

Oxidizing Agents: Improve overall strength of the dough by reinforcing the proteins. They contribute to the finished crumb and make the dough easier to shape. Available in powder or tablet form; added early in the mixing process.

Reducing Agents: Reduce the overall strength of the dough by weakening the proteins, which reduces mixing times and machinability.

Mold Inhibitors: Compounds that slow down the growth of mold and other bacteria in bread. These include natural food products like sorbic acid, vinegar, raisin juice concentrate, and whey protein, or through the addition of nutrients like calcium, sodium propionate, or potassium sorbate.

Protease Enzymes: Naturally produced by yeast; some bakers elect to add more protease enzymes to improve machinability and mixing time by weakening the proteins present in the dough. Available in powder or tablet form, and usually added early in the mixing process.

Calcium Peroxide: An oxidizing agent that is also used to dry out a dough, removing excess stickiness and making it easier to shape. It is available in powdered form and should be mixed in with the other dry ingredients.

Crumb Softeners: A type of emulsifier added to the dough to bond to the starches in the flour and produce a lighter finished loaf.

Dough Strengtheners: A type of emulsifer that bonds to protein and improves the overall strength of the gluten.

INCLUSIONS AND GARNISHES

Additional ingredients are sometimes added to bread doughs to provide flavor and texture to the finished loaf. While these additions can feasibly be any number of ingredients, the most common in this context are cheese, herbs, spices, chocolate, extracts, nuts, seeds, and dried fruit. It is important to note that inclusions with high levels of salt, acid, or sugar can affect gluten development and should be added to the dough at the end of mixing to prevent damage to the gluten strands. In addition, inclusions with high levels of moisture can alter formula ratios. It is important to test formulas when introducing new inclusions, as they can affect the end result. It is also advisable to consider flavor intensity when choosing either inclusions or garnishes; overly intense flavors can mask the natural flavors of the bread. Similarly, mild flavors may need to be added in higher quantities to ensure their flavor is properly imparted into the dough. It is also important to consider that the size of the garnish will break down during mixing.

Choosing Inclusions and Garnishes

There are several factors to consider when adding an inclusion to a bread dough or a garnish to a finished loaf or roll. First and foremost should be the quality of the ingredient being added. The process of bread baking is a complex one, with flavor, texture, and appearance affected by every step of the process. Using high-quality ingredients is a must to maintain the integrity of the finished product. In addition, bakers should ask themselves the following:

- Does the inclusion/garnish complement the bread? For example, are the ingredients traditional to the region where the bread originated? Regionality and authenticity should be considered, as they set the stage for classic flavor pairings.
- Is the size of the inclusion/garnish appropriate for the finished product? A loaf may contain cubed cheese, which will then be visible when the loaf is cut into and will impact the flavor of each bite. A roll, on the other hand, may be garnished with shredded cheese so that it is visible to the customer and is also part of each bite.
- What is the best way to incorporate the inclusion/garnish? Should the ingredient be mixed directly into the dough (see below), or should it be sprinkled on top of the shaped loaf or roll?
- How will the inclusion/garnish affect the final product? Is the ingredient salty or sweet? Is the texture chewy or crisp? Does the ingredient contain high levels of moisture or acid? When incorporating an ingredient as a garnish on top of the baked loaf or roll, the question of when to add it becomes important. For example, when adding cheese as a topping, it should be added part way through the baking process; if it is added too early, it will burn and develop a bitter flavor. In these cases, the size of the bread and the baking time will determine when to add the topping.
- What is the purpose of the inclusion or garnish being added to the dough? In general, inclusions should be visible to the customer. They should contribute flavor, appearance, and texture to the final product.

When to Add Inclusions and Garnishes

Every inclusion or garnish is different and therefore should be treated as such: bakers should analyze each ingredient and how it will affect the bread before deciding how to incorporate it and in what quantity. However, there are some general guidelines to consider.

- Most inclusions should be added after the dough is fully developed. They should be added into the dough and incorporated on first speed to prevent the ingredients from being broken down. Still, it is important that the ingredient is fully incorporated into the dough. Generally speaking, most garnish ingredients should be added later in the mixing. If added too early, they might get shredded or broken down into too small a particle size by the repeated action of the mixer.

- Regardless of the ingredient, inclusions should be added gently on very low speed. It's almost like massaging them into the dough, and it may be necessary to finish the incorporation of inclusions by hand to preserve their integrity. Properly executed, inclusions should be fully incorporated into the dough and visible to the eye (meaning they have had the proper length of mixing time). Inclusions should not be severely broken down or unevenly mixed (a sign of too long and too little mixing time, respectively).
- Garnishes added to the exterior are most often added to the dough once it has been shaped. This may happen before or after final proof has taken place. Common exterior garnishes include grains, nuts, seeds, oats, and cheese.

Cheese

Cheese can add a richness in flavor and texture to the finished breads. Depending on the type of cheese being used, it's especially important to consider factors such as flavor intensity (strong cheeses, like blue cheese, may impart particularly intense flavors), saltiness, and moisture content.

All cheeses should be incorporated during the last few minutes of mixing. Their soft texture ensures that they will most likely be broken down, not necessarily visible in large clumps. Soft cheeses greatly influence the fat and salt content of the dough. The additional fat provides flavor and tenderness. Salt levels may vary, and salt amounts in the base formula may need to be adjusted to compensate.

Semisoft cheeses, such as Gouda, can vary greatly in the correct time to be added. Essentially, that depends on how finely chopped or grated the cheese is. Grated cheese can be added earlier in the mixing process to ensure full and even incorporation. Cubed or chopped cheese should be approximately the size of a sugar cube, and should be added at the end of mixing to prevent the cheese from losing its shape. Pain au Fromage (page 227) is an ideal way to utilize leftover cheese. The bread can be made with any cheese, or any combination, but as previously mentioned; the formula may need some basic adjusting to ensure proper flavor and texture. The formula can be standardized by using the same variety of cheese in the preparation each time.

Hard cheeses such as Parmesan are usually grated and added during mixing or as a garnish on top of the dough. Grated hard cheeses can be added up to 20 percent of the weight of the flour during mixing (do not wait until the end, as the cheese will become an essential part of the dough's structure and texture).

Herbs

Herbs add a fresh, bright note to doughs, and are commonly used as garnishes in a variety of breads and rolls. Make sure the herb being used complements the flavor of the dough, and especially any other inclusions that may be added as well (such as garlic, olives, or nuts). Because of their delicate texture, herbs should be added at the very end of mixing, to prevent bruising the leaves or overly breaking them down. Herbs should not be used as a garnish before baking—herbs applied to the exterior of the dough will burn in the high oven temperatures. Instead, breads can be brushed with oil and garnished with fresh herbs after baking. This also adds an incredible aroma to the finished loaf.

Spices and Seeds

Spices can be added during mixing or as a garnish on top of breads. Whole spices and seeds, such as fennel, cardamom, and coriander, should be lightly toasted before being ground and added to doughs. The exception to this rule is when they are added as a garnish to the exterior of dough—then they will toast properly on the outside of the dough from the oven heat. The spices listed above are common flavorings for many old-fashioned breads, specifically traditional rye breads. The amount used varies with the formula, but their flavor should be mellow so as not to overpower the dough. Other spices, such as cinnamon, cloves, allspice, and nutmeg, are often added to enriched or sweet doughs.

Seeds can be used to create soakers (see page 131), or they can be mixed in as an inclusion, or used as a garnish. Seeds should not be added early in the mixing process—they should be reserved and added at least halfway through the mixing. It is important that the dough develops a base level of structure before adding seeds, which can otherwise cut through weaker gluten strands and/or prevent them from properly forming. Common seeds used in bread baking are poppy seeds, the aforementioned fennel and caraway seeds, pumpkin seeds, sesame seeds, sunflower seeds, and flaxseed.

Chocolate

In this text, semisweet or bittersweet chocolates are the most common used; other varieties can be too sweet, don't give the same depth of flavor, and will burn more easily owing to the high milk-solid and sugar content. Chocolate can be added in small or large chunks, but should always be added at the end of the mixing time. If it is added at the beginning, the friction of the mixer may cause the chocolate to melt. Cocoa powder is most often added for color, and should be added during the earlier stages of mixing to ensure even incorporation. Cocoa powder should be used instead of shaved chocolate, which will not add enough flavor or color to the dough to justify its use. Avoid having pieces of chocolate exposed on the exterior of a shaped loaf or roll—those pieces will burn during baking and contribute a bitter flavor to the finished bread.

Extracts and Other Flavorings

Extracts are most commonly added to enriched doughs, though there are a few traditional loaves made in certain regions of the world that are the exception to that rule. Generally speaking, they should be added at the beginning of mixing time to ensure even incorporation. The same can be said for vanilla. Both vanilla beans and vanilla extracts are commonly used to flavor fillings and toppings for enriched breads as well.

Other ingredients, such as garlic, onions, olives, citrus zest, corn kernels, sun-dried tomatoes, or cooked potatoes are all common additions to bread dough. When to add these ingredients depends largely on the desired finished texture. For example, if whole or half olives are desired, they should be added at the end of mixing time. If an olive-speckled dough is desired, they can be added earlier to create a different end result. Remember to consider that some items can benefit from being added earlier in the mixing process. Sun-dried tomatoes, for example,

Rehydrating

Many dried fruits benefit from being rehydrated before being added to a bread dough. The ideal way to rehydrate, or plump, dried fruit is to soak it in hot, flavorful liquid, until it absorbs some of the liquid, usually doubling in size. Fruit can be rehydrated by being warmed with enough flavorful liquid (fruit juice, liquor, vinegar, etc.) to cover, and then placed over medium heat until the liquid begins to simmer. Once the liquid is at a full simmer, the pot should be removed from the heat and allowed to sit for 4 to 6 minutes, until the fruit has doubled in size and has absorbed some of the liquid.

Particularly small or delicate dried fruits can be handled slightly more carefully by placing them in a heat-safe bowl. Bring the soaking liquid to a simmer, and pour over the dried fruit, then allow time for plumping to occur. Some formulas may call for the fruit to be drained of excess liquid, but generally speaking, the formula can be altered to compensate for the addition of this liquid, which imparts excellent flavor to the dough.

can rehydrate during mixing owing to the liquid added to the dough. Citrus zest can release some of its essential oils as it is broken down during mixing. Other ingredients, like corn kernels, should be added at the end of mixing to preserve their appearance and structure.

Because the moisture in fresh fruits varies and could drastically affect a ratio of bread dough, most formulas that use fruit inclusions employ dried fruit. Some dried fruits, like currants, can be added whole to bread doughs. Others should be chopped before being added, but take care not to chop them too fine. The bits of dried fruit inside a dough add wonderful visual appeal, and that can be lost if the fruit breaks down too much during mixing.

Some formulas may call for the fruit to be rehydrated, or plumped, before being added to the dough. This is especially true of dried fruit with a brittle or especially chewy texture. Fruits should be soaked in a flavorful liquid, such as fruit juice or liquor; the flavor is absorbed during the soaking process and then imparted to the dough. Take care not to oversoak the fruit, as this can cause the fruit to break down excessively in the mixer. It is not always necessary to soak dried fruits; some have enough moisture (even in their dehydrated state) to be added to the dough as is, which eliminates the need to make adjustments to the moisture content of the base formula.

Nuts

Nuts should most often be lightly toasted before being added to doughs as an inclusion. The exception, again, is if the nuts are being used as a garnish. Take care not to overtoast the nuts, as this will make them bitter. Some nuts can drastically affect the composition and appearance of the dough. Walnuts, for example, can turn a dough purple if they are mixed for too long, as their skins contain a purplish pigment that is activated with excessive mixing, changing the pH of the dough.

In general, whole nuts should be added at the end of the mixing time. If chopped nuts are being added, it comes down to personal preference. Nuts can be fully incorporated into the dough by adding them halfway through mixing time or left in larger pieces for a textural element in the finished bread.

Grains

While grains can be added to doughs in a variety of ways (particularly ground into flours and mixed in a traditional fashion), they can also be added to create textural interest and nutritional value. Two methods can be used to add grains to doughs (similar methods are also used for seeds, or for grains and seeds in combination): sprouting and soakers (see sidebars, pages 75 and 76).

It is important to note that most whole grains require the use of a soaker. It is possible to buy mixes of grains and seeds that can be used to create soakers, but creating your own mixtures allows for more creativity and flexibility in formula development. Consider the role of cost control and time management when deciding whether to purchase grain mixes or create your own. Grains that are commonly added to bread doughs include oats, millet, wheat berries, cracked wheat, rye berries, cracked rye, hemp, and spelt.

Making a Soaker

Both grains and seeds benefit from being soaked before their incorporation into a dough. This is for a number of reasons, but the main one is to hydrate and soften the grain. Another reason is that these ingredients are very dry and can create textural problems in the dough, absorbing moisture from the liquid in the formula. Their shapes can also be damaging to the formation of gluten strands. Their hard exterior and firm shape can actually cut through weak, not fully formed gluten strands and prevent them from gaining the proper levels of strength.

There are two types of soakers: hot soakers and cold soakers.

Hot soakers are produced by bringing liquid to a boil, pouring it over the grain, and stirring to ensure all the grains are in full contact with the water. The grains then soak in the hot liquid for a period of time, typically overnight at room temperature, before being added to the dough. Hot soakers work by pre-gelatinizing the starch of the grain. This can be beneficial to the final dough because it can improve the exterior crust formation and also decrease overall baking time. However, hot soakers are also said to impart less flavor to the final dough.

Cold soakers are produced by combining the grains with room-temperature liquid and soaking overnight. Cold soakers need a minimum of an overnight soak to fully soften the seeds (hot soakers can be used in as little as 1 hour after their production). However, cold soakers are said to have increased flavor.

Soakers should be added to the dough after it has started to develop gluten, except when you need the water from the soaker to hydrate the dough. The notable exception to this rule is rye doughs, which have minimal gluten development, so soakers should be added at the beginning of the mixing time. The soaker should be added and incorporated on medium speed until fully incorporated.

Sprouting Grains

Sprouting grains provides a number of benefits. The process softens the grain and begins germination. This amplifies the presence of natural enzymes and vitamins, offering additional health benefits. Sprouting also changes the structure of the carbohydrates inside the grain's makeup. These carbohydrates are converted into vegetable sugars, rather than fibrous starches that are more difficult to digest. Finally, sprouting reduces the presence of antinutrients that are naturally present in the grain, which can prevent the body from fully absorbing the beneficial nutrients during digestion.

After the grains are sprouted, they can be ground into flour. Sprouting takes additional preparation, but it is a remarkably simple process. Sprouted grain flours are also available for purchase, but can be relatively expensive per pound. Here's how to sprout the grains:

- Rinse the grains thoroughly. Transfer to a storage container, preferably with a cover (though you can cover the container with plastic wrap during storage if need be).
- Heat water to the appropriate temperature (the recommended range is between 95 and 105°F/35 and 41°C).
- Soak the grains for the required time. Some grains require an overnight soaking, while some may require 2 days. Store the grains at room temperature while they are soaking.
- After the correct length of sprouting time, drain the grains for 30 minutes. Rinse the drained grains under warm water for 10 minutes.
- Drain the grains, then spread them onto a sheet tray to dry for 2 hours. The dried sprouted grains can be used or ground into flour.

MAKING ARTISAN BREADS

4

Artisan bread requires just a few key ingredients, but it is the ratio of those ingredients and the way they are handled that produce a wide array of finished loaves. Proper execution comes from a detailed understanding of each step in the process—what needs to happen to produce a properly mixed dough, how to determine if a dough is properly proofed, the ideal oven conditions and temperature, and so on. This chapter will discuss steps in the bread-making process, along with information on two important tools in bread baking: bakers' percentage and desired dough temperature (DDT). This information will provide bakers with an in-depth understanding of how to properly execute the formulas given in the remaining chapters of this book.

BAKERS' PERCENTAGE

The bakers' percentage is used in all types of baking, but it is especially useful in bread baking. The concept is simple, an invaluable tool that helps bakers create new formulas and adjust current formulas. Used properly, bakers' percentage can lead to consistency in production, better calculation of the absorption rate of the flour, and the ability to increase or decrease formula yields, as well as making it easier to compare formulas and offering a way to check a formula for balance and potential defects.

The bakers' percentage is always based on the total weight of the flour in the formula, which is always represented by the value of 100 percent. All other ingredients in the formula are calculated in relation to the flour, each assigned a percentage of its own based on the amount of it in the formula. Thus, the total percentage in the formula adds up to over 100 percent; this is of no consequence. The purpose of the bakers' percentage is to provide a way to calculate and/or adjust the other ingredients in the recipe.

Generally speaking, the bakers' percentage is used only with weight measurements, not volume. All the ingredients are listed in similar units of measure. That is, all ingredients are listed in pounds or grams. Also, it is possible for ingredients to be more than 100 percent in the formula, a reflection of the dough in question. For example, some very hydrated doughs might have water listed at 125 percent. If the baker finds the dough to be overhydrated, he or she could alter the formula by adjusting the percentage of water.

Example 1: Calculating the Bakers' Percentage

INGREDIENT	US	METRIC
Flour	2 lb 3¼ oz	1 kg
Water	1 lb 1¾ oz	500 g
Salt	1¾ oz	50 g
Yeast	⅓ oz	10 g

As the base of the formula, the bakers' percentage for the flour is 100 percent. The baker can then calculate the bakers' percentage for the ingredients in one of the following ways:

Division: dividing the weight of the water by the weight of the flour gives the bakers' percentage as 50 percent. For example,

1 lb 1¾ oz/500 g (water)/2 lb 3¼ oz/1 kg (flour) = 0.50, or 50%

Cross Multiplication: multiplying the weight of the water by 100, then dividing by the weight of the flour gives the bakers' percentage as 50 percent. For example,

1 lb 1¾ oz/500 g (water) × 100 = 50,000/2 lb 3¼ oz/1 kg (flour)
= 0.50, or 50%

If the same calculations are applied to the entire formula, the completed recipe would read as follows:

INGREDIENT	US	METRIC	BAKERS' PERCENTAGE
Flour	2 lb 3¼ oz	1 kg	100%
Water	1 lb 1¾ oz	500 g	50%
Salt	1¾ oz	50 g	5%
Yeast	⅓ oz	10 g	1%
Total	3 lb 7 oz	1.6 kg	156%

Example 2: Calculating Ingredient Amounts Using Bakers' Percentage

The reverse method can also be used to determine the amounts for the ingredients in a formula, based on simply the percentages. This is especially useful for bakers looking to increase or decrease a formula to produce a specific yield. Say, for example, a baker wants to increase the example recipe to yield 18 lb 11¾ oz/8.5 kg dough. Here's how it is done:

Division: dividing the percentage for the flour (100%) by the total percentage for the full formula yields how much of the total formula is the flour. This number can then be multiplied by the desired yield to determine the correct flour amount.

100 (flour %)/156 (total %) = 0.64 × 18 lb 11¾ oz/8.5 kg (desired yield)
= 12 lb ¼ oz/5.5 kg flour

Cross Multiplication: multiplying the percentage of the flour by the desired yield, then dividing by the total percentage of the formula yields the amount of flour needed in the recipe.

100 (flour %) × 18 lb 11¾ oz/8.5 kg (desired yield)
= 850,000 g/156 (total %) = 12 lb ¼ oz/5.5 kg

If the same calculations are applied to the entire formula, the completed recipe would read as follows:

INGREDIENT	US	METRIC	BAKERS' PERCENTAGE
Flour	12 lb ¼ oz	5.5 kg	100%
Water	6 lb ¼ oz	2.7 kg	50%
Salt	9¾ oz	275 g	5%
Yeast	2 oz	55 g	1%
Total	18 lb 11¾ oz	8.5 kg	156%

Example 3: Calculating Preferment Quantities Using Bakers' Percentage

Bakers may choose to alter a formula by adding a preferment to it, and the bakers' percentage makes this easier to do. For example, suppose a baker wants to add a preferment to the example dough using 20 percent of the flour weight. The preferment will also contain 64 percent water and 1 percent yeast in proportion to the total flour in the formula.

First, the baker calculates the amount of flour to use in the preferment by multiplying the weight of the flour by the 20 percent.

12 lb ¼ oz/5.5 kg (flour weight) × .20 = 2 lb 6½ oz/1.1 kg

Next, the baker determines the weight of the water and yeast in the formula, utilizing the preferment flour amount and the respective water and yeast percentages.

2 lb 6½ oz/1.1 kg (preferment flour weight) × .64 (preferment water %)
= 1 lb 9½ oz/698 g (preferment water amount)

1,090 (preferment flour weight) × .01 (preferment yeast %)
= ⅓ oz/11 g (preferment yeast amount)

Therefore, the finished preferment formula would read as follows:

INGREDIENT	US	METRIC
Flour	2 lb 6½ oz	1.1 kg
Water	1 lb 9½ oz	698 g
Yeast	⅓ oz	11 g
Total	3 lb 15½ oz	1.8 kg

To adjust the final recipe, the baker subtracts the quantities of ingredients used in the preferment from the corresponding initial amounts listed in the formula. Because there was no salt added to the preferment in this example, the salt quantity in the recipe is not altered.

Flour: 12 lb ¼ oz/5.5 kg (original formula) – 2 lb 6½ oz/1.1 kg (preferment)

= 9 lb 9¾ oz/4.4 kg (adjusted formula)

Water: 6 lb ¼ oz/2.7 kg (original formula) – 698 g (preferment)

= 4 lb 7¾ oz/2 kg (adjusted formula)

Yeast: 2 oz/55 g (original formula) – ⅓ oz/11 g (preferment)

= 1⅔ oz/44 g (adjusted formula)

INGREDIENT	US	METRIC	BAKERS' PERCENTAGE
Flour	9 lb 9¾oz	4.4 kg	100%
Water	4 lb 7¾oz	2 kg	50%
Salt	9¾ oz	273 g	5%
Yeast	1⅔ oz	44 g	1%
Total	13 lb 12½ oz	6.3 kg	156%

Finally, by applying the methods from example 1, the baker can determine the bakers' percentages for the final preferment in relation to the full formula. The finished formula would read as follows:

INGREDIENT	US	METRIC	BAKERS' PERCENTAGE
Preferment			
Flour	2 lb 6½ oz	1.1 kg	25.0%
Water	1 lb 9½ oz	698 g	16.0%
Yeast	⅓ oz	11 g	0.1%
Dough			
Flour	9 lb 9¾ oz	4.4 kg	100.0%
Water	4 lb 7¾ oz	2 kg	50.0%
Salt	9¾ oz	273 g	5.0%
Yeast	1⅔ oz	44 g	1.0%
Preferment	3 lb 15½ oz	1.8 kg	41.0%
Total	18 lb 11¾ oz	8.5 kg	197.0%

DESIRED DOUGH TEMPERATURE (DDT)

Understanding the desired dough temperature (DDT) is vital to successful mixing of any bread dough. The temperature of the finished dough directly affects the rate of fermentation, so bakers control this fermentation rate by controlling the dough temperature. For proper fermentation, the ideal DDT for most doughs is between 74° and 82°F/23° and 28°C. Any cooler than this, and the dough will ferment very slowly, possibly affecting production times. If the dough is warmer than this, it will ferment very quickly, which can affect the flavor of the finished dough, as well as its strength.

There are five factors that contribute to the dough's final temperature:

- Temperature of the room
- Temperature of the flour
- Temperature of the water
- Mixer friction (heat generated from the mixing process)
- Temperature of the preferment (if applicable)

Note: Another factor that can affect the final dough temperature is the addition in a large amount of eggs, butter, a soaker (see page 131), and the like.

Using DDT to Calculate Water Temperature

The baker can control only some of the factors affecting final dough temperature. The room temperature, for example, can be adjusted generally, but it is likely established by the time of mixing. The flour temperature could in theory be controlled, but generally flour is stored and used at room temperature. The mixer friction is a set number, but varies from mixer to mixer. The temperature of the preferment, if being used, can be controlled somewhat (the water and flour temperature when it is mixed, and its storage temperature), but its temperature at the time of mixing cannot be adjusted, so it is also a fixed factor.

The only factor at the time of mixing that can be controlled and adjusted by the baker is the water temperature. Using the desired dough temperature and the other temperatures listed above, a baker can determine the temperature of the water that should be added to the dough during mixing.

Example 1: Calculating Water Temperature When Not Adding a Preferment

DDT = 76°F/24°C

Room Temperature = 68°F/20°C

Flour Temperature = 68°F/ 20°C

Mixer Friction = 8° (See Note)

First, the base temperature is calculated. The base temperature is found by multiplying the DDT by the number of factors affecting the final temperature. In this case, there are three factors: room temperature, flour temperature, and mixer friction.

$$76°F/24°C \times 3 = 226°F/108°C$$

Then, the known temperatures are subtracted from the base temperature to determine the water temperature.

$$226°F/108°C - (68 + 68 + 8) = 82°F/28°C \text{ (water temperature)}$$

Example 2: Calculating Water Temperature When Adding a Preferment

DDT = 77°F/25°C

Room Temperature = 65°F/18°C

Flour Temperature = 65°F/18°C

Preferment Temperature = 69°F/21°C

Mixer Friction = 10° (See Note)

First, the base temperature is calculated. The base temperature is found by multiplying the DDT by the number of factors affecting the final temperature. In this case, there are four factors: room temperature, flour temperature, preferment temperature, and mixer friction.

$$77°F/25°C \times 4 = 308°F/154°C$$

Then, the known temperatures are subtracted from the base temperature to determine the water temperature.

$$308°F/154°C - (65 + 65 + 69 + 10) = 99°F/37°C \text{ (water temperature)}$$

Note: The mixer friction is not an actual temperature (therefore not listed in Fahrenheit or Celsius). It is simply the logged increase in temperature during mixing.

Calculating the Mixer Friction

To calculate the mixer friction, the baker must make an educated guess in regard to the water temperature (typically between 65 and 80°F/18 and 27°C) used to hydrate the dough. When all the other known temperature factors have been recorded, the dough can be mixed. If the baker uses the temperature of the fully mixed dough as the final dough temperature (FDT), the math can be reversed using the FDT and the water temperature—hence, the difference is the mixer friction.

THE TWELVE STEPS OF BREAD BAKING

The process of bread baking can best be understood by examining the twelve steps for producing bread. Each step plays an important role in the final result. They are:

1. Scaling
2. Mixing
3. Bulk fermentation
4. Folding
5. Dividing
6. Preshaping
7. Bench rest
8. Shaping
9. Final fermentation
10. Scoring
11. Baking
12. Cooling

Understanding each of these steps involves a detailed look at their contributions to the finished loaf, and this understanding can then aid in troubleshooting or perfecting the process.

Step 1: Scaling

Without consistent measure of ingredients, a baker's batches of bread will lack consistency. Therefore, the importance of proper scaling cannot be overstated. As mentioned in Chapter 2, scales (either digital or balance beam) are the ideal form for the production baker to measure ingredients and doughs.

The scale should be properly zeroed (including any vessel that is being used to measure the ingredient) before any measuring is begun (see page 85). Unless otherwise specified by the formula, it is best to keep ingredients separate until the time of mixing. For example, neither salt nor sugar should come in direct contact with the yeast. Instant yeast can be mixed directly into the scaled flour (if it is not being used to make a sponge), helping to disperse it and prevent direct contact. This also helps incorporate the yeast more efficiently, allowing it the best amount of time to dissolve and rehydrate for ideal fermentation.

Accurate and careful scaling is the foundation to good bread baking. Keep ingredients organized and separate until mixing.

Remember to plan ahead—take the temperature of ingredients like preferments and flour before moving forward. If the ingredients are scaled ahead of time, label them clearly and remember to account for their temperature during storage, as that might affect the DDT.

Step 2: Mixing

So much of the bread-baking process is determined or begun during mixing. Characteristics that define the finished loaf, such as gluten development, interior structure, dough temperature (affecting fermentation), and the overall consistency of the dough, are all connected to the mixing process.

MIXING METHODS

Different breads require different methods of mixing, though those methods are similar, with just a few key alterations. Doughs that require very little gluten development (such as certain flatbreads) are mixed minimally, often by hand—this is called a *short* mix.

Doughs needing more gluten development (such as baguette dough) are mixed more substantially; this is called *improved* gluten development. Other doughs (such as bagel dough) require substantial mixing to fully develop the gluten strands—this is called *intense* gluten development.

SHORT MIXING METHOD Short mixing is a method that produces minimal gluten development. The mixing can be done by hand but also can be recreated using a mixer; the latter is ideal for larger quantities in a production setting. The ingredients are combined in the mixer bowl on first speed, and continue to be mixed on first speed until they obtain some gluten development. The mixing then ceases, preventing the dough from achieving further gluten development. The dough then needs a long fermentation time (with two to four folds spread throughout the total time) to further develop the overall strength of the dough.

Considerations:

- A smaller amount of yeast should be used in the formula to compensate for the longer fermentation time.

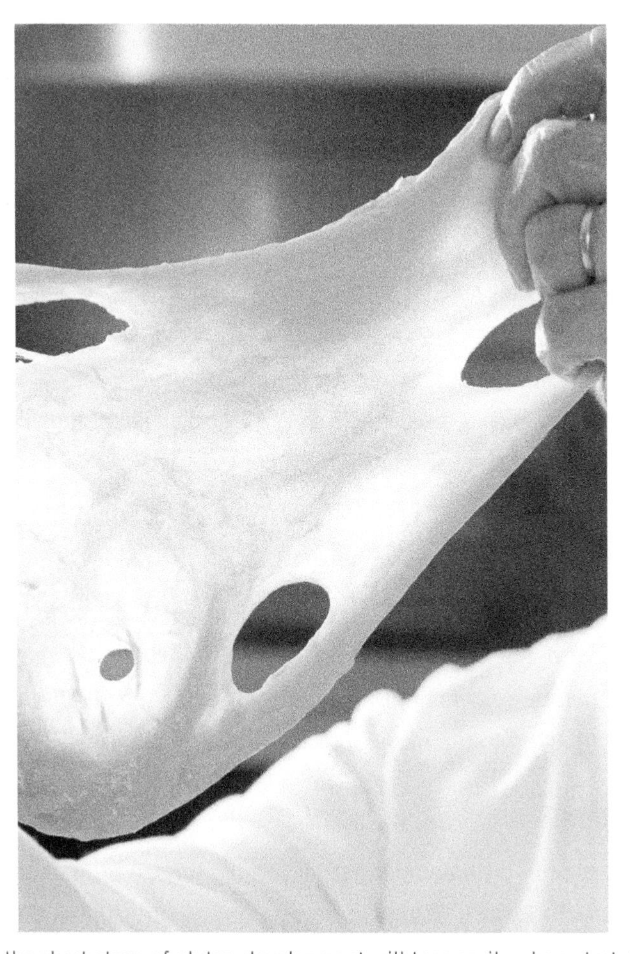

Dough mixed to the short stage of gluten development will tear easily when stretched.

- Doughs mixed using the short method should be soft—better extensibility is ideal for repeated, easy folding.
- Short mixed doughs should still be easy to shape, though in general they are very soft and full of large gas bubbles.
- Dough is barely oxidized owing to the minimal mixing, meaning the interior crumb color will be light and creamy in the finished loaf.
- Crumb structure will be light, very open, and relatively irregular (because of the size and number of gas bubbles created by long fermentation). Final volume will suffer because there is less structure in the dough overall.
- The flavor of the finished dough might be slightly acidic as a result of extended fermentation.

IMPROVED MIXING METHOD The improved mixing method is a middle ground between short mixing and the full gluten development achieved by intense mixing. Ingredients are mixed to combine on first speed, and then the dough is mixed on second speed

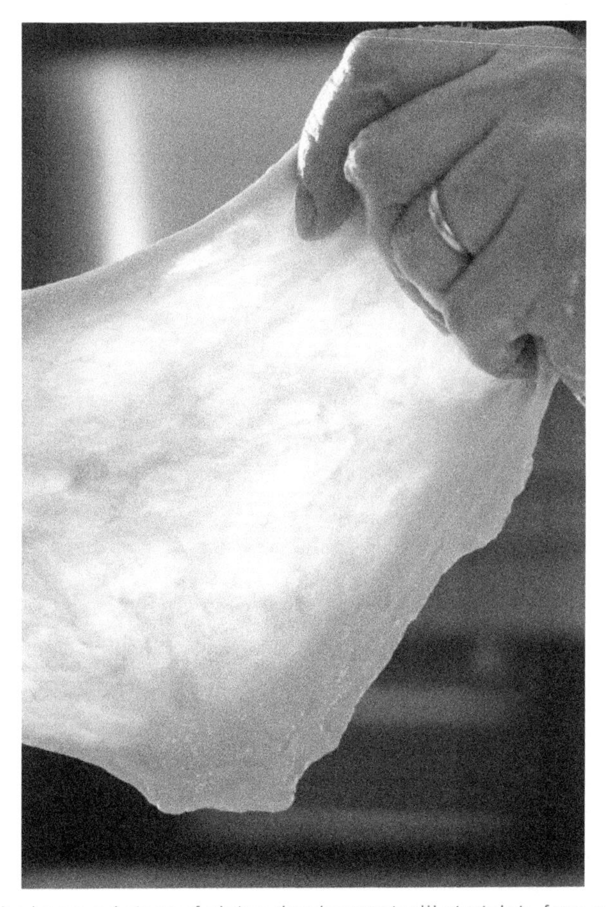

Dough mixed to the improved stage of gluten development will stretch to form a thin membrane that does not tear.

to half gluten development. The result is a dough that requires less fermentation and folding overall, and has more structure during shaping and baking.

Considerations:

- Doughs mixed using the improved mixing method can be slightly soft or slightly firmer—as long as they are extensible enough to handle the required fermentation times.
- Longer fermentation times than doughs mixed with the intense method provide excellent pronounced flavor.
- Crumb color will be relatively light and creamy, with a slightly tighter crumb structure that still has some variance and openness due to gas bubbles.
- The finished loaf will have more substantial volume than doughs mixed using the short method.

INTENSE MIXING METHOD The intense mixing method produces a very stiff, firm dough with full gluten development. Ingredients are incorporated on first speed until combined, and then the dough is mixed to maximum gluten development on second

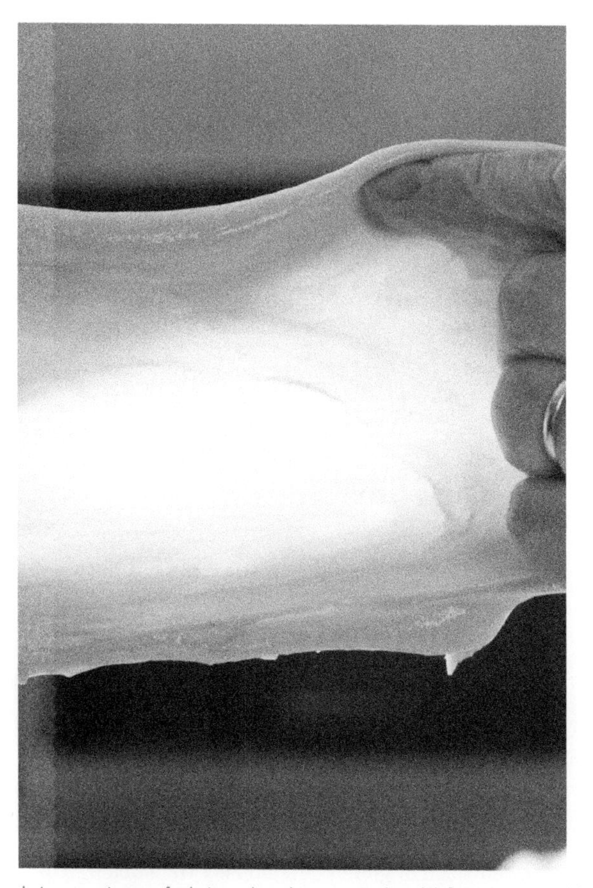

Dough mixed to the intense stage of gluten development should form a very thin membrane when stretched. It should be transparent enough to see light through it.

speed. The dough is very strong and requires minimal fermentation. In fact, too much will produce excess strength and negatively affect the dough.

Considerations:

- This mixing method is fast and efficient, and the dough is easily handled. It can stand up to a heavily mechanized process for shaping.
- The extended mixing time creates substantial oxidation, producing a white crumb color.
- The crumb structure of the finished loaf will be very regulated and very tight with an even grain.
- Shelf life is diminished owing to a shorter fermentation time.
- Fully developed gluten structure allows for maximum gas retention, leading to excellent final volume.

Examples of Doughs Mixed to Different Gluten Strengths

SHORT MIX (WHEN MIXED BY HAND)	IMPROVED MIX	INTENSE MIX
Ciabatta	Semolina bread	Bagels
Focaccia	Focaccia	Pretzels
Irish soda bread	Ciabatta	Bialys
Sourdough bread	Pane di como Filone Lean dough	Challah Brioche

EFFECT ON DOUGH	SHORT MIXING	IMPROVED MIXING	INTENSE MIXING
Consistency	Fairly soft	Medium	Stiff
Strength	Lacking	Slightly lacking	Good to strong
First fermentation	Long	Medium	Short
Machining	Difficult	Possible	Ideal
Final proof	Short	Medium	Long
EFFECT ON BREAD			
Volume	Small	Medium	Large
Color	Creamy	Less creamy	White
Crumb	Open, irregular	Open, irregular	Tight, regular
Flavor	Very complex	Complex	More bland
Shelf Life	Longer	Medium	Shorter

Calculating Mixing Time

To help professional bakers determine the appropriate mixing times, classifications have been developed for each type of mixing method. The classification is based on the number of revolutions of the mixer per minute and how many total revolutions are needed for each type of mixing method. The total number of revolutions does not include the first stage of mixing, during which the ingredients are combined and fully incorporated. This initial stage generally takes 4 to 5 minutes.

- Short mixing method: 600 revolutions total (on first speed, after ingredient incorporation)
- Improved mixing method: 1,000 revolutions once the speed has been raised to second speed (after ingredient incorporation).
- Intense mixing method: 1,600 revolutions once the speed has been raised to second speed (after ingredient incorporation).

Begin by checking the manual of the mixer being used to determine the number of revolutions it makes in a 1-minute period. Then apply the logged number to the following equation:

Total revolutions required by mixing method / revolutions per minute (RPM) = total mixing time

Note: Using this method cannot substitute for a baker's experience or developed sense from working with the dough.

No matter which mixing method is chosen, the first step in mixing is to calculate the water temperature needed (see page 82) to achieve the desired dough temperature (DDT). Once the mixing is begun, the flour is hydrated, which initiates the formation of gluten in the dough, known as dough development.

- **Pickup:** The dough is just beginning to come together and begins to resemble a sticky mass.
- **Clean-up stage:** The dough begins to pull away from the sides. This is the last point at which hydration can be adjusted.
- **Preliminary development:** The dough begins to form a fully homogenous mass and starts gluten development.
- **Initial development:** The mixer speed is raised to continue gluten development.
- **Final development:** The dough is fully homogenous and very elastic—this happens by maintaining the raised speed of initial development for as long as it takes to reach the desired result.

The longer the dough is mixed, the more structured and organized these gluten formations become, making the dough more extensible and elastic. In addition, oxygen is incorporated into the dough, which strengthens the gluten. The incorporation of too much oxygen can damage the color and flavor of the dough. The dough continues to be mixed until it reaches the desired stage of development (as specified in the formula).

Pickup Stage

Clean-Up and Preliminary Development Stage

Initial Development Stage

Final Development Stage

As soon as mixing is completed, it is important to check the temperature of the dough to see if the proper desired dough temperature (DDT) has been reached; the proper salt content should also be checked. If it has, the baker can proceed with the next step in the bread-baking process, which is bulk fermentation. If the dough is too cold or too warm, the first fermentation time will need to be altered to compensate and allow for proper fermentation. Do not continue mixing the dough if the temperature of the dough is too low. While mixer friction will raise the temperature of the dough slightly, mixing will also continue to develop the gluten and create a tough, overdeveloped dough.

Double Hydration Mixing

Very hydrated doughs, such as ciabatta, are often mixed using the double hydration method. This technique incorporates the water in two stages. First, some water is added at the beginning of the mixing to create a medium-soft consistency, hydrate the dough, and help with gluten formation and development. When the dough has reached nearly full gluten development, the rest of the water (10 to 15%) is added little by little until it is fully incorporated into the dough. This technique produces a soft and strong dough.

Effects of Mixing

Both physical and chemical changes occur in the dough during mixing. The physical changes begin as soon as the flour and water come into contact with one another. The water hydrates the flour, immediately affecting the starch and protein present in it. Flour has two types of starch: native starch and damaged starch. Native starch can only absorb the water on its surface, while damaged starch absorbs nearly its own weight in water. The proteins in the flour can absorb 200 to 250 percent of their weight in water. These proteins inflate and become attracted to one another, and this begins the process of forming gluten bonds. Starch absorbs water faster than protein, a factor that affects the mixing process. Initially, doughs are mixed on first speed to develop the gluten; then, the mixer speed is increased to organize the developed gluten bonds. However, if the speed is raised too quickly, the gluten will not be fully developed as it begins to get organized with the higher speed of kneading, and thus there will not be full, proper development.

Once the formation of the gluten has begun, the mixing works the gluten into an organized structure. Though it was traditionally done with hands kneading the dough, this mixing is now done with the dough hook attachment on the mixer, which stretches the gluten bonds with the first movement and then folds them over one another with the second movement. The gluten chains become longer and more developed as the dough mixes, forming more and more structure. Note that if the dough is undermixed, the

Mixing Log						
Dough Name	**Mixing Method**	**Room Temperature**	**Flour Temperature**	**Preferment Temperature**	**Water Temperature**	**Final Dough Temperature**
Lean Dough	Improved	60°F	60°F	68°F	48°F	78°F
Semolina	Intense	60°F	60°F	65°F	53°F	82°F
Focaccia	Short	60°F	60°F	62°F	49°F	70°F

A mixing log is a helpful tool for tracking the progress of a dough during the mixing process.

gluten will not develop enough to have the proper structure; if the dough is overmixed, the gluten will develop too much and eventually the bonds will break.

Adding water to the dough also introduces chemical changes. The first two chemical reactions most affecting the dough are fermentation activity and enzyme activity. The quantity of water added drastically affects the rate of these reactions. For example, a very hydrated dough will generate faster fermentation, so lower quantities of yeast should be used.

Another important chemical change that occurs is oxidation. As mentioned previously, some oxidation will always occur, and to some extent it can lend positive results. For example, the presence of oxygen helps to form stronger gluten bonds. Oxygen also helps to form the crumb structure of the finished loaf by introducing micro cells of air. This air provides room for the gas cells produced by fermenting yeast to accumulate, forming the "alveoles" of the crumb.

However, overoxidation will negatively affect the carotenoid pigments in the flour. These pigments are responsible for the creamy color of the flour and also for the bread's pleasant aroma. Too much oxygen will deteriorate these pigments, giving the finished loaf a lighter color and also affecting its flavor. To slow down the negative effects of oxidization, the baker can use an alternative mixing method called autolyse.

Autolyse

Autolyse is an alternative mixing method by which salt is *not* added during the initial stages of mixing and there is overall a shorter mixing time. Omitting the salt at this point slows down the chemical reactions and reduces the action of the proteases in the flour, allowing greater hydration (3 to 5% more). This process increases the dough's extensibility, thereby decreasing the strength of the dough. The finished dough is more fully hydrated and the finished loaf has greater volume. Traditionally, this method also held back the yeast during the early stages of mixing, so as to prevent the fermentation process from beginning and to keep acidity out of the dough as it is mixed. However, if dried instant yeast is used, it is added in the early stages of mixing to ensure that it has enough time to be incorporated and hydrated.

Factors Affecting Mixing Time

- **Type of mixer:** Motor speed, the shape of the hook, and mixer design can all affect the stretch and fold motions of the hook against the dough. For example, a bread made with a spiral mixer requires only 5 minutes of mixing on second speed, while the same bread made with an oblique mixer requires 12½ minutes. However, exact times always depend on the make and model of the mixer used.
- **Batch size:** The amount of dough in relation to the size of the mixing bowl is crucial. Mixers are generally designed to perform optimally at full capacity. Generally, smaller batches mix faster because they come in contact with the hook more often. In these instances, the mixing time should be reduced to compensate, though this must be done by experience—there is no precise formula to calculate these alterations.
- **Type and quality of flour:** Varying flours absorb water and develop gluten differently.
- **Absorption of water and hydration of the flour:** Lower hydration creates a stiffer dough with less extensible gluten. These doughs require a longer mixing time to ensure proper development.
- **Incorporation of enrichments and inclusions:** See the section below for more information on the incorporation of additional ingredients.

Incorporation of Enrichments During Mixing

Enrichments and inclusions are not always added to bread doughs, but they can affect the mixing process when they are added at that point. Understanding when and how to incorporate these ingredients can prevent problems with mixing enriched doughs.

FAT

Smaller quantities of solid fat (2 to 4%) can be incorporated with the flour and liquid at the beginning of the mixing. Larger quantities (5 to 15%) should be incorporated when the dough is halfway through its development. If larger quantities are added too early, the fat could lubricate the proteins in the flour, preventing the formation of gluten and the bonding of the strands. Any quantity greater than 15 percent should be incorporated at the very end of mixing, when the gluten is almost fully developed. This ensures that the dough is strong enough to tolerate a large quantity of fat.

Liquid fats should be added during the initial stages of mixing to ensure proper incorporation. Very large quantities can be added using the double hydration method of mixing (see page 92), but it is important to add the fat slowly to ensure proper incorporation into the dough.

SUGAR

Small quantities of sugar (up to 12%) can be incorporated into the dough at the beginning of mixing. Larger quantities (up to 19%) should be incorporated in several steps. Because sugar is hygroscopic, it could absorb too much water if added to the dough all at once at the beginning of mixing, thus negatively affecting the overall hydration of the dough, as well as the formation of gluten structure. High levels of sugar (20 to 30%) should be added at the end of mixing, when the gluten is almost fully developed.

EGGS

Eggs should be incorporated at the beginning of the mixing, as they aid in hydrating the flour. Always add some quantity of water (at least 10%), even if eggs are the primary hydration ingredient. The final product will then have a lighter and moister crumb.

DRY INCLUSIONS

Additional dry enrichments such as malt or milk powder should be incorporated at the beginning of the mixing, along with the flour and water.

SOLID INCLUSIONS

Any coarse ingredient that won't dissolve fully into the dough should be incorporated at the end of the mixing time. Once gluten has been properly developed, the mixer is turned back to first speed, and the inclusions are added and mixed just until fully incorporated. Gentle incorporation on first speed ensures that the ingredients remain intact within the dough and also prevents damage to the gluten structure.

STEP 3: BULK FERMENTATION

After the dough has been properly mixed, bulk fermentation begins, a step so named because the dough ferments in one large mass directly out of the mixer. Fermentation is defined as the breakdown of compound molecules in an organic substance under the effect of yeast or bacteria. Lactic fermentation is used to make cheese, butter, and yogurt. Acetic fermentation is used to produce vinegar. Alcoholic fermentation is used to make wine, beer, and bread. In bread baking, fermentation occurs by the conversion of sugars into alcohol and carbon dioxide as a result of yeast activity and naturally occurring bacteria.

Dough before bulk fermentation.

Dough at the end of bulk fermentation.

What Is Fermentation?

Wheat flour contains various types of glucides that are used throughout fermentation. Some are used as is, while others are degraded by naturally occurring enzymes to aid in the fermentation process. Glucose and fructose are the simplest of the glucides and make up about 0.5 percent of the flour. They are directly consumed by the yeast by penetrating the outer membrane of the cell. These simple sugars are then transformed into alcohol and carbon dioxide, thus beginning the fermentation process. Their consumption takes place in the first 30 minutes after mixing.

More complex glucides are also present in the flour, including saccharose and maltose (making up about 1% of the flour). They are more complex, and require time to be broken down into a simpler sugar and then consumed by yeast. Saccharase and maltase, which are enzymes present in yeast (see page 61), break down these two glucides into glucose, which is consumable for the yeast.

Starch is another complex glucide, which comprises about 70 percent of the flour. There are two types of starch in most flours: amylose and amylopectin. These starches require multiple steps to be broken down into yeast-digestible glucose. Amylose is first broken down by the enzyme beta amylase into maltose, and then the maltose is broken down by maltase into glucose. Amylopectin is degraded by the alpha amylase enzyme into dextrin, and dextrin is then further degraded by the beta amylase enzyme into maltose. Finally, maltase breaks it down into glucose.

As previously discussed, the bulk of the starches used during fermentation are those that are degraded during the milling process, which are more capable of absorbing water during mixing. This moisture absorption is what triggers enzyme activity, beginning the process of fermentation.

Enzymatic Balance in Flours

The alpha and beta amylase enzymes that naturally occur in flour can vary owing to the germination stage of the wheat. When the wheat plant is ready to sprout, the germ sends enzymes into the endosperm. These enzymes transform the complex components of the endosperm into smaller nutrients used by the germ. Storage of harvested wheat is easier for farmers if the wheat is harvested before it begins to sprout. Therefore, the final milled flour often lacks these enzymes.

Millers compensate for lesser quantities of enzymes by adding malted flour and/or fungal enzymes during the milling process to properly balance the flour so that fermentation can occur.

Effects of Fermentation

Fermentation affects the dough in several ways. First, the dough rises and gains volume from the production of carbon dioxide. In the initial stages of fermentation, the gas is dissolved in the water present in the dough that is not bonded to the flour. After the water in the dough becomes saturated with gas, it begins to cause internal pressure. This pressure stretches the gluten in the dough. Owing to the structure formed during mixing, the gluten withstands this process and retains the carbon dioxide, causing the dough to rise.

The dough also begins to become acidic from the production of organic acids. These acids lower the pH of the dough and are a sign of proper fermentation activity. The production of acids also improves the shelf life of the finished product by slowing the staling process and preventing mold growth. The acidity also helps develop aroma and flavor in the dough. Some acids are developed by the production of alcohol during the fermentation process, while others are simply naturally occurring organic acids created by secondary reactions in the dough. The first fermentation (bulk fermentation) is the building block for these complex flavors and aromas. Longer bulk fermentation times increase the gluten development and strength of the dough, consequently reducing the extensibility and increasing the elasticity of the dough.

Factors Affecting Fermentation

Fermentation is affected by a variety of factors in relation to the dough. The first is temperature. Yeast activity is faster at higher temperatures and slower at lower temperatures. Optimal fermentation temperature is around 76°F/24°C. Another important factor is the quantity of yeast in the dough. To control fermentation and allow for enough time to benefit the process, the quantity of yeast should be controlled and limited (anywhere from 0.5 to 2%).

The quantity of salt and sugar impact fermentation. Salt slows fermentation activity; optimal fermentation can occur when salt is around 2 percent. Sugar can increase or slow fermentation depending on its quantity. Small amounts of sugar (around 5%) increase fermentation, because in small quantities sugar serves essentially as an additional food source for yeast. A larger amount of sugar (12% or more) slows the fermentation, because the yeast cells function differently with a large quantity of sugar present in their environment. Commercial yeast works best when the dough has a pH of between 4 and 6. The lower the pH, the more fermentation is reduced.

About Bulk Fermentation

The first fermentation is crucial because the dough is fermenting in one mass, developing most of its final flavor and aromatic characteristics. The fermentation helps the dough rise, develop optimal flavor, and increase the bread's shelf life. In general, it is best to allow (both in terms of formula development and mixing time) for a long bulk fermentation period. If, for production purposes, a long first fermentation is not possible, bakers can choose to use a preferment to increase the benefits of fermentation in a shorter period of time (see page 120).

Often times, recipes call for the dough to be folded one or more times during the bulk fermentation process (see the following section for more information). The dough is gently stretched to elongate it, and then it is folded into thirds. Lean doughs, such as baguette, are generally folded one or more times to improve structure and expel gas. Some enriched doughs are not folded so as to avoid overdevelopment, as they have been extensively mixed. Rye doughs are also generally not folded so as to prevent the dough from collapsing.

It is important to remember that each step of bread baking affects the other steps. For example, longer bulk ferment times require shorter mixing times; otherwise, the dough would become overly structured between the strengthening of gluten bonds during mixing and the acidity production of fermentation. Doughs mixed by hand are inherently less developed, and therefore require a longer fermentation time to ensure the proper structure is formed.

STEP 4: FOLDING

Folding is a simple, but important step in the bread-making process. The dough is folded after bulk fermentation to redistribute the available food supply for the yeast, equalize the temperature of the dough, expel the built-up gasses and alcohol (by-products of fermentation), and further develop the gluten. Some doughs are folded more than once, adding a second fold after the final fermentation.

Doughs that have a typical level of hydration (around 67% or less) should be treated carefully during the folding process. It is more difficult for the carbon dioxide produced during fermentation to leaven the bread, owing to the density of the dough and the tightness of the gluten structure.

Doughs that have a high level of hydration, such as ciabatta, require more aggressive treatment during folding. It is especially important for very wet doughs to develop more gluten structure during folding. This ensures that the dough will maintain its inner structure and hold onto its shape during the remaining steps of production.

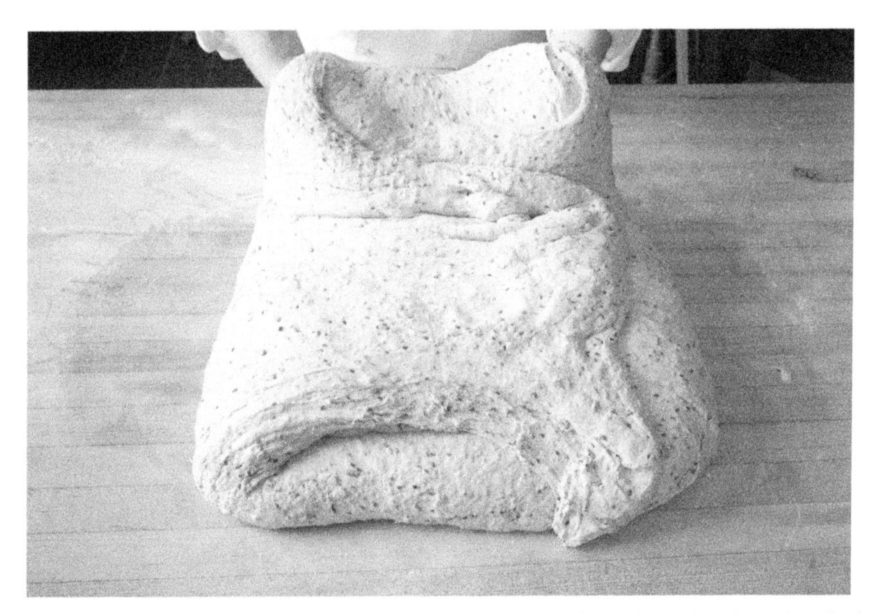

Dough is folded once in both directions to ensure that all of the dough has been handled.

STEP 5: DIVIDING

Dividing is the term used to describe the period after bulk fermentation during which the dough is divided into desired portions. This can happen either manually or by machine, but either technique must be done carefully to avoid damaging the dough's gluten structure. It is also important to note that the dough contains a lot of gas, which can make it slightly more difficult to handle at this point.

If the dough is being divided by hand, it is best to cut accurately. The goal is to have a single piece of dough rather than several pieces balled together. If the dividing is being done by machine, the type of machine used is important in making sure the dough maintains its structure. For example, a divider that uses suction is not ideal for a very gassy dough. Generally speaking, hydraulic dividers are best for doughs with long bulk-fermentation times. If the dough is overly handled or otherwise damaged during division, it can inhibit the next step of the process, known as preshaping.

Dough should be divided quickly to prevent it from deflating.

STEP 6: PRESHAPING

Preshaping is the term used to describe the gentle reforming of the dough that has been divided. The shape is loose, generally a ball or cylinder depending on the intended shape of the final loaf. This preshaping allows the dough to ferment briefly in close to its final shape, easing the final shaping process.

By this stage in the process, the dough should have the proper levels of strength and is in the final stages on its way to baking. However, if the dough is wrong in some way, preshaping is an opportunity to make alterations. For example, if the dough is lacking structural integrity, it can be preshaped more tightly to increase the gluten strands one more time. Or, if the dough has become overly extensible, it can be very loosely shaped to prevent further or excessively agitating the structure.

Some very delicate loaves, such as ciabatta (which is highly hydrated), do not need preshaping. Also, during preshaping it is advisable not to use an excessive amount of flour; a lightly dusted work surface is acceptable.

STEP 7: BENCH REST

The bench rest, also known as intermediate fermentation, is a short period of time when the gluten strands relax before the dough is shaped. The preshaped loaves are covered and allowed to rest for 10 to 20 minutes, depending on the formula's guidelines.

Cover the dough during the bench rest to keep it from drying out and forming a skin.

STEP 8: SHAPING

After resting, the dough is given its final shape, either by hand or by machine. Shaping should be done on a lightly floured surface to avoid incorporating excess flour into the dough. Additionally, the dough should be minimally worked to achieve the desired shape, avoiding any damage to the structure. The final shaping is the last stage at which alterations can be made to the dough if it is overly extensible or underdeveloped.

Miche

A miche is a large rustic loaf traditionally made in the French countryside from whole wheat flour. Weighing 8 to 12 lb/3.6 to 5.4 kg, a miche can feed an entire family for a week or more. The loaves are made very large to prevent the bread's drying out too quickly. The bread comes from a time when families didn't have the opportunity to bake their bread every day, so they baked larger loaves with a longer shelf life. The size of the loaf lends itself to decorative scoring or stenciling.

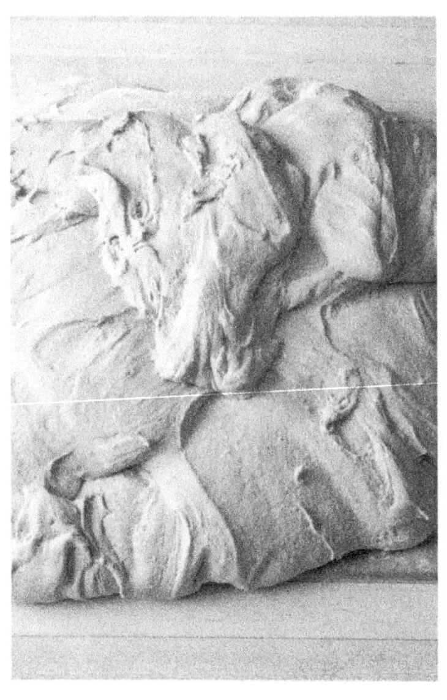

A miche is traditionally made with a hearty, whole wheat dough.

Preshape and shape a miche like a smaller boule. It should be a large, tight round.

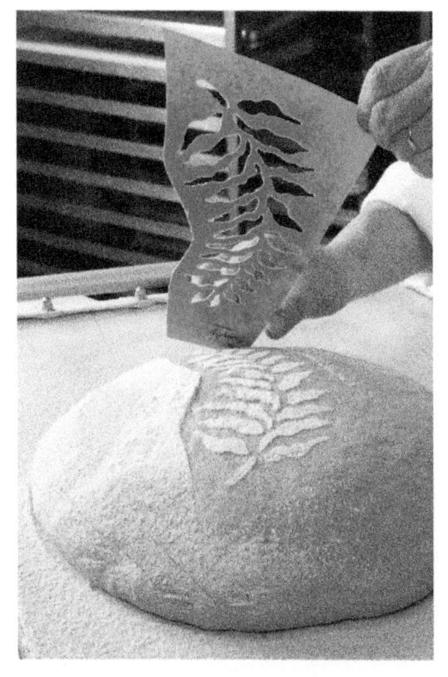

Because of its size, a miche is an excellent vehicle for stencils or decorative dough.

Slice the miche as it is needed, and it will keep for 1 to 2 weeks.

Shaping a Baguette

A baguette is an elongated loaf. To form a baguette, the baker begins with an oblong preshaped piece of dough. It is placed lengthwise on a lightly floured work surface, and

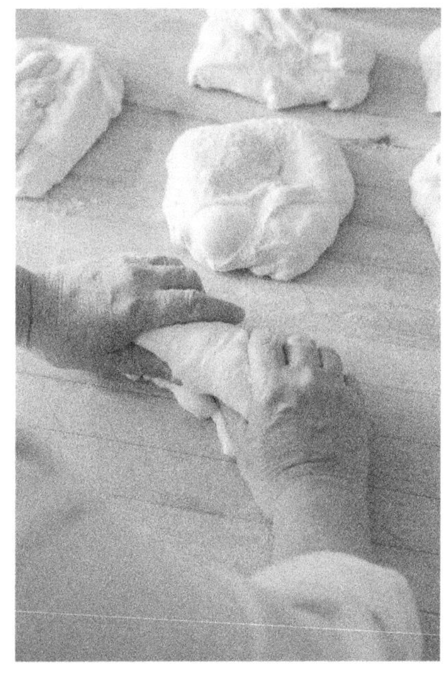

To preshape the baguette, fold the left and right sides of the dough toward the center, being careful not to de-gas the dough. Fold the top of the dough to meet the bottom, and seal the seam with the heel of your hand.

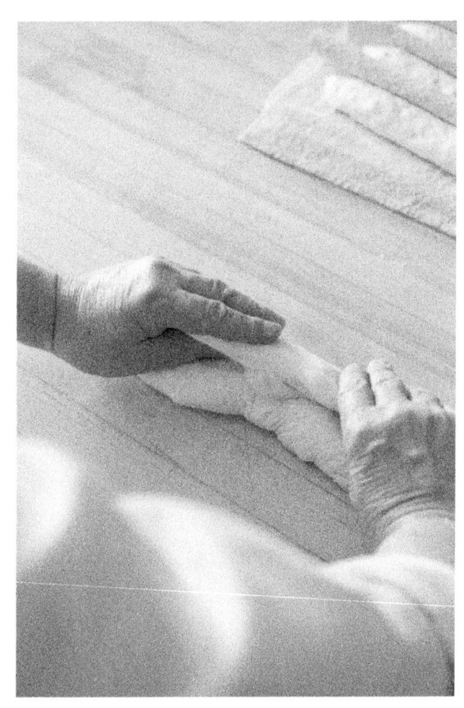

To shape the baguette, fold the dough over your right-hand thumb to create a seam.

Use the heel of your left hand to seal the seam around your thumb. Repeat this step twice, for a total of three times.

Start in the center of the baguette and use both hands to roll the dough until it is 21 in/53 cm long and slightly tapered.

the top piece of the dough is folded inward toward the center, with the baker pressing with the heel of his hand to seal it. Care must be taken not to degas the dough too much. The dough continues to be folded until it is folded over itself completely and sealed. The baker places the dough, seam side down, on the work surface and begins rolling with the fingertips, starting from the inside and working out to lightly taper the edges. This is repeated until the dough is 21 in/53 cm in length. The shaped loaves are gently placed into a couche for final fermentation.

Shaping a Boule

A boule is a rounded loaf. To form a boule, the baker begins with a preshaped round piece of dough. She folds the dough in half, then rotates it slightly, and repeats with another fold. This process expels some of the gas present in the preshaped loaf. The dough is then placed, seam side down, on a lightly floured surface, and the shape is tightened by rotating the dough between cupped hands in a downward circular motion.

Preshape the dough into a round by rolling it with the palm of your hand against the table.

The final boule should be much tighter than the preshape. Pull the round toward you, using the friction from the table to tighten the dough.

This is continued until the dough has reached the desired tightness and is in a uniformly rounded shape. The baker gently places the boules in baskets or couches for final fermentation.

Shaping a Bâtard

A bâtard is an oblong loaf. To form a bâtard, the baker begins with an oblong preshaped piece of dough. He places it lengthwise on a lightly floured work surface, and folds the top piece of the dough inward toward the center, pressing with the heel of his hand to expel gas and seal. This is repeated until the dough has been folded over itself completely and is sealed. The baker places the dough, seam side down, on the work surface and begins rolling with the palm of his hands, starting from the inside and working out to lightly taper the edges. He then gently places the shaped loaves into a couche for final fermentation.

Fold the top piece of the dough toward the center and press with the heel of your hand, repeating until the dough is completely folded over itself.

Using your thumbs as guide, roll the dough with the palms of your hands, starting from the center. The ends should be slightly tapered.

STEP 9: FINAL FERMENTATION

The last stage of fermentation occurs when the shaped bread is left to ferment one additional time, either at room temperature or in a proof box, which provides the ideal warm and humid environment. Additional carbon dioxide will form inside the shaped dough, giving the bread a final rise and creating additional volume for a light and airy texture. The dough should be kept away from cold air or drafts to prevent its drying out.

Dough that has been shaped by machine will more readily withstand a longer final proof time compared to dough that has been hand shaped. That is, dough that has been shaped tightly will need a longer final fermentation time to reach the proper texture for baking. Hand-shaped dough is generally looser in structure and has a higher gas content, which means it requires less final proofing time.

For final fermentation, or proofing, the dough may be placed on floured boards or in lined bannetons.

Some doughs will rise dramatically during proofing, but for others, it may be difficult to judge visually.

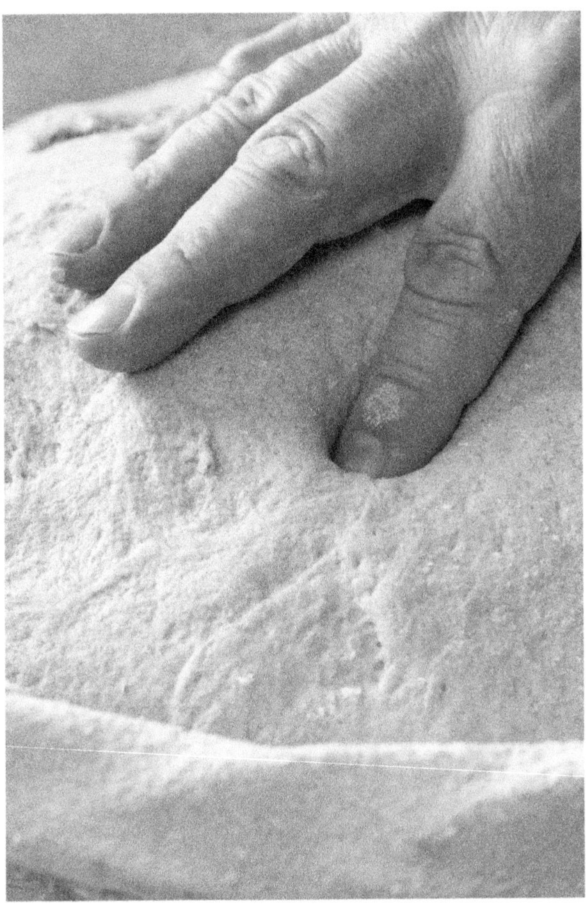

To test if a dough has finished proofing, lightly press a finger into the dough.

If the dough is properly proofed, your finger will leave a mark that should retreat about halfway.

STEP 10: SCORING

Scoring is the process of cutting the dough (with a razor blade, lame, knife, or scissors) just before baking. The result is both decorative and practical—the vent created by scoring allows a place for the crust to break, so that the expanded loaf remains evenly shaped. The process also allows the bread to rise more during the first stage of baking, also known as oven spring, thereby preventing the finished loaf from being overly dense. Very hydrated and/or enriched doughs generally do not require scoring.

Although doughs should be perfectly proofed when baked, adjustments in the scoring can accommodate under- or overproofed dough so that the final result is still acceptable. If dough is underproofed, it should be given a deeper score to encourage better expansion in the oven. Likewise, an overproofed loaf should be given a shallow score to prevent deflation from overexpansion.

Score quickly with very sharp tools to prevent damaging the dough.

Examples of decorative scoring.

Baguettes are scored with horizontal, elongated scores that overlap.

The method in which the scoring is done will change the appearance of the finished loaf. In fact, scoring is often used as a way to identify different loaves of bread in a bakery.

Only very sharp tools should be used to score the dough—dull blades tear through the skin on the dough's surface and can damage the integrity of the dough. Scoring tools can be stored in a cup of water near the oven. This is ideal both from a sanitation perspective and for ease of use—a moist blade cuts more easily through the surface of the dough.

This baguette was baked with the seam on the side of the dough instead of on the bottom.

The scores on this baguette are irregular and not consistently placed.

The scores on this baguette don't overlap, while the overlap is an identifiable trademark of a baguette.

This baguette on the right is too short and unevenly rolled.

This baguette is underbaked with an undeveloped crust.

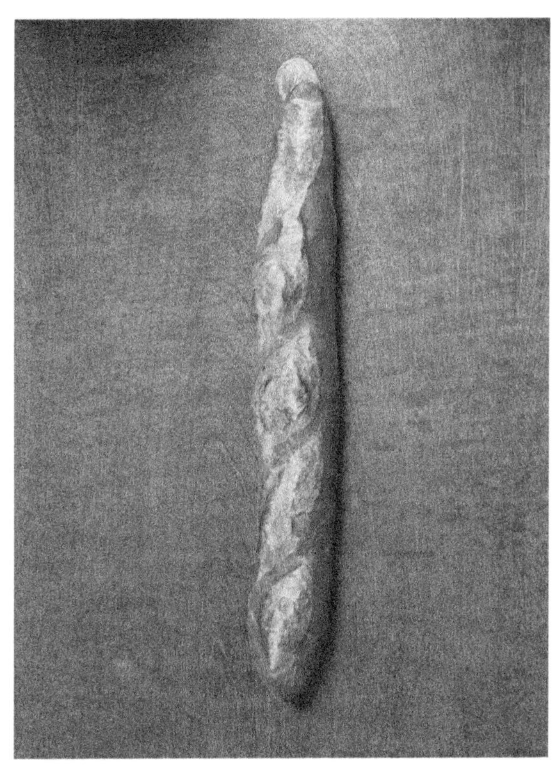

This baguette is overbaked. The crust will be hard and tough, and it may taste burned.

Scoring requires a precise, dexterous, and light movement. Most important, it takes a skilled hand and extensive practice to learn well. There are a large variety of scoring styles, reflecting the shape of the bread and the traditions of individual loaves.

For elongated loaves, the classic (or baguette) score is often used. This technique involves holding the blade at a 45-degree angle over the surface of the dough to create a horizontal, repeating incision. The score is either done in one long motion, from one end of the loaf to the other, or in several repeating, shorter scores. Repeating scores should overlap slightly (by about one-third the length of the score). Only a thin film of dough is cut and forms an "ear" over the cut opening. If the angle is not correct, the score can be too deep, causing excessive spreading of the dough during baking. If the incision is correct, however, the sides of the loaf will spread slowly and the layer of dough created by the incision will, to some extent, temporarily protect the surface from the heat, encouraging a better rise during oven spring.

Another cut for elongated loaves is the sausage cut (see page 112), commonly used for bâtards. In this technique, the blade is held perpendicular to the dough. The cuts are made diagonally, almost perpendicular to the side of the loaf. The number of cuts can vary, but the space between each cut should not be more than ½ in/1 cm.

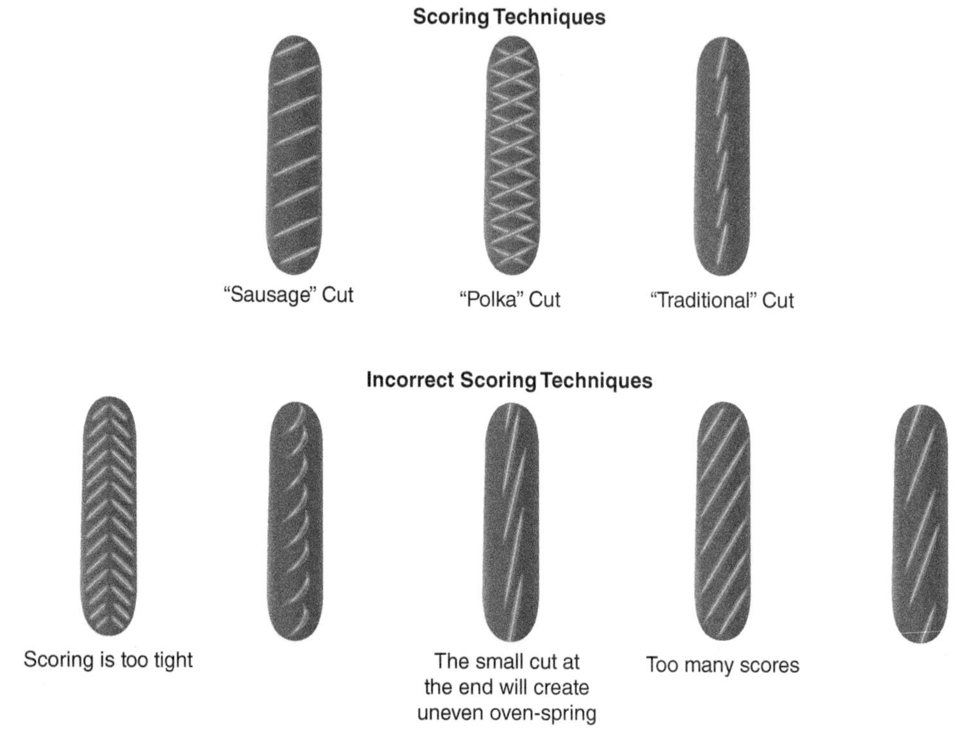

Elongated loaf scoring techniques.

Another score commonly used for bâtards is the polka dot, or criss-cross, score. The blade is held perpendicular to the dough and cuts are made diagonally, starting from one side of the loaf and moving across it. Then, additional cuts are made at an opposite angle using the same movement. The number of scores will vary depending on the size of the loaf, but they should cover the full surface of the dough.

For other loaves, such as boules, the scoring possibilities are limitless. Leaf cuts, diamond cuts, wave cuts, and so on are just some used by bakers to vent and identify their breads. Whichever styles are chosen, it is important to stay consistent. This makes it easier for staff to sell the bread, as well as for customers to identify the products being sold.

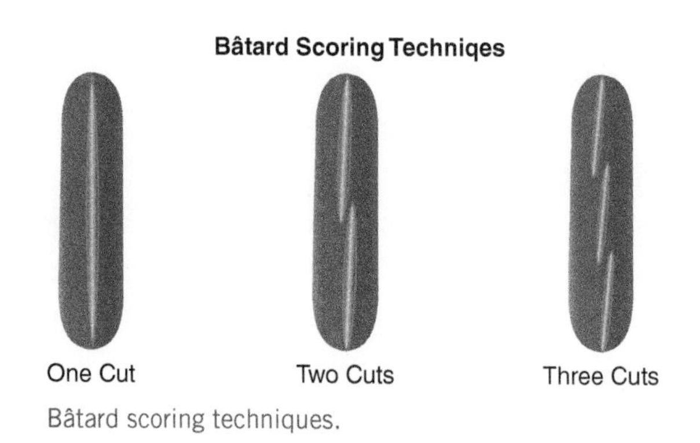

Bâtard scoring techniques.

Boules Scoring Techniques

"Straight Line" Cut

"Square" Cut

"Cross" Cut

"Diamond" Cut

Boule scoring techniques.

In addition to scoring, a decorative stencil can be applied to dough before baking.

STEP 11: BAKING

After the dough is scored, the loaves are loaded into the oven. This can be done manually or automatically. In either case, care should be taken to ensure the dough is not mishandled or abused in a way that could affect the structure. When the loaves are transferred to the oven, for instance, they must be given the proper amount of space to encourage optimal heat circulation and result in an ideal finished product.

Dough being loaded using an automatic dough loader, which uses a rotating linen to deposit the dough directly onto the oven surface.

Oven Temperature

There are no standard suggested oven temperatures for baking bread. Rather, the temperatures vary drastically depending on the formula and the desired end result (as well as the type of oven). However, some generalizations can be made.

Lean doughs should be baked around 480°F/248°C in a deck oven, or around 450°F/232°C in a convection rack oven. In general, a 12 oz/340 g baguette should bake in 20 to 24 minutes. The baguettes should be vented in the last 5 to 6 minutes of baking. If they bake in less time, the oven temperature is too hot and should be adjusted. Inversely, longer baking times indicate that the oven temperature is too low and should be raised.

In general, lean doughs should have a baking time that is three times longer than their venting time. So, if a baguette bakes for 18 minutes, it should be vented for 6 minutes and have a total baking time of 24 minutes. This can vary widely depending on the dough, however, so rely on your experience to know when the breads are properly baked.

Once an ideal temperature has been determined, remember that larger pieces of dough will need a longer baking time at a lower temperature, and smaller pieces of dough will benefit from baking at a higher temperature for less time. For rye breads, a high temperature must be used at the beginning of the baking process, and then the temperature should be lowered after the bread has been loaded, with the vent opened 5 minutes into the baking. Rye breads require a long baking time (about 1 hour per 2 lb 3¼ oz/1 kg of dough). For more information on rye breads, see pages 342–377.

Steaming

After the bread is loaded into the oven, it is ideal (in most cases) to steam the dough. Steaming affects the bread in several ways. For one, the steam creates moisture that adheres to the dough's surface inside the hot oven. This moisture makes the surface more extensible and able to develop an even oven spring with the extensive pressure from gas at the beginning of the baking period. This creates a loaf with better volume. Additionally, the moisture on the surface delays the process of exterior evaporation, thereby delaying formation of the crust, which in turn leads to a thinner and crisper end result. Finally, the moisture from the steam generates a slight dilution of the starches present on the surface of the dough. When these starches gelatinize, the crust gains a deeper and glossier color.

Steam should be injected into the oven before and after the loaves have been loaded. This way, the loaves are loaded into a moist environment, and they remain in that environment for as long as possible. Convection rack ovens do not allow steaming before loading, but their steam generators are more powerful and can sufficiently steam after loading. The amount of steam should be enough to form a light film of water on the surface of the dough. Because ovens vary drastically, it is impossible to list suggested steam times. This must be done by sight and experience of the baker.

Baking

As soon as the heat reaches the dough and begins to increase its temperature, a series of chemical and physical reactions begin to take place. During the first 4 to 6 minutes, the yeast and enzyme activity are stimulated by the quick increase in temperature. Because more sugar is transformed by the enzymes and because the yeast's metabolism is kick-started, a large amount of carbon dioxide will be produced and retained by the gluten structure, thereby developing additional volume in the bread. This is referred to as oven spring.

After oven spring occurs, the reactions can be monitored based on the bread's temperature.

122°F/50°C: The starch granules start to swell and the yeast starts to die.

140°F/60°C: The starch begins to gelatinize—the granules burst and release long chains of starch that form a complex gelatinous matrix. After cooling, this becomes the bread's crust.

145°F/62°C: The yeast cells die and end all yeast activity. However, continued gas production still causes additional increases in volume.

153°F/67°C: The gelatinization of starches is complete.

165°F/73°C: The gluten begins to coagulate and the protein chains begin to solidify. Once they do, the structure of the loaf is fully set.

180°F/82°C: All enzymatic activity has come to a halt and no further chemical reactions occur in the dough.

212°F/100°C: Evaporation on the surface begins to create the final crust. The final colorization of the crust happens at higher temperatures, when the sugars in the dough begin to caramelize (known as the Maillard reaction).

Determining Doneness

The doneness of bread is often difficult to determine. Visual components are important, but they are not enough. Bread can brown fully if baked at a very high temperature, even if the inside structure is not properly set. Other guidelines for doneness include time, along with the experience of the baker, to determine how much time each loaf generally takes in the oven. Sound is another excellent doneness-determining factor. Finished loaves have a hollow sound when tapped with a finger after baking. A properly baked loaf will lose 10 to 20 percent of its original weight after baking, though this number depends on the type of bread being baked. The greater the surface area of the bread being baked, the more weight loss will occur.

Another technique is to gently press the side of the bread after it is removed from the oven. A crust that is too soft will reveal an insufficient baking time. Remember, though, that the crust can become soft and soggy as the bread cools down. This often happens to doughs made with flour containing a high ash content, such as whole wheat or rye flour. The bran present in the flour can absorb a large quantity of water, and if it is not allowed to fully evaporate during baking, the crust will not set properly. To prevent this, the bread should be allowed to fully dry before being removed from the oven. This can be done by venting the oven (or simply opening the oven door for a few minutes) so as to remove any steam that has yet to evaporate inside the oven.

STEP 12: COOLING

Bread must be properly cooled to set the final structure. See the chart on page 89 for examples of interior colors and crumb structures of finished breads. During cooling, some moisture is released, generating a loss of weight in the final product (as mentioned above). This moisture, if not allowed to dissipate into the air, will condense on the bread's surface and create a soggy texture when it is reabsorbed into the crust.

The pressure inside the loaf will equalize as the bread cools. That is, during baking, the heat causes the gas to expand. When the gas is expelled from the loaf, some air will replace the gas. Cool air occupies less volume inside the finished loaf compared to

Place the finished bread on a rack to cool.

the expanded carbon dioxide. This can cause the bread to contract and shrink slightly. Indeed, the crumb retracts easily because of its elasticity. The firm outer crust, however, audibly cracks in places. Crusts that don't crack may be soggy after cooling.

Aroma is also distributed during the cooling process. Despite the pleasant aroma of freshly baked bread, the bread should be cooled completely before being consumed. The aroma from the crumb will diffuse into the crust, creating a fuller and more complete aromatic experience.

The Process of Becoming Stale

As soon as bread has cooled, the degradation process begins, commonly known as staling. Staling occurs in three ways: the crumb, the crust, and the aromas.

Crumb: The crust begins to degrade owing mostly to the migration of water. When the dough is mixed, water surrounds the starch particles. During starch gelatinization in the baking process, the water moves from the outside of the particle to the inside. This bursts the starch particles and extends some of the chains of starch. As the bread cools, these chains retract to their original position, making the crumb more dense and forcing it to lose its softness.

Crust: During staling, the crust loses its shine and becomes duller. The crispness changes, becoming harder in dry weather and soggier in humid weather.

Aroma: Most aromas dissipate into the air shortly after the bread has cooled. More and more aroma is lost as the bread stales. The staling process of the crumb and crust reduce their aromatic qualities as well, lessening the quality of the finished loaf.

How to Delay Staling

- Dough with higher hydration levels will stale more slowly.
- Longer fermentation times produce more acidity, which reduces staling.
- Maintaining proper volume during baking ensures the product will dry out less quickly.
- Proper air circulation during cooling ensures ideal and maximum shelf life.
- Freezing can delay staling. In fact at very low temperatures (–4 to –22°F/–20 to –30°C), staling is almost stopped entirely. The more quickly bread is frozen after it has been baked and cooled, the longer it can survive in the freezer. Frozen breads can be defrosted in a 400°F/204°C oven (steam is recommended) for 4 to 5 minutes, then allowed to finish defrosting at room temperature.

It is important to note that as the bread cools down, the bakery warms up. This can affect the desired dough temperature for those breads being mixed, as well as their fermentation and proofing times. Keep track of these temperatures and compensate as necessary.

LEAN DOUGH BREADS AND ROLLS

5

The phrase "lean dough" describes a dough that is made without enrichments such as fats or sweeteners, and is instead made only with flour, water, yeast, and salt. Generally speaking, the resulting loaves have crusty exteriors, as is the case for a baguette or ciabatta. Some lean doughs are handled in ways that give them a chewier texture, such as pretzels or bialys. Some gain their flavor through inclusions or toppings, such as focaccia, while others acquire the bulk of their flavor through the use of preferments, like sourdough bread. In addition to flavor considerations, preferments give the professional baker more flexibility with the bakeshop schedule. Preferments shorten the fermentation times for doughs without sacrificing flavor, so the mixing and baking times can be adjusted to accommodate doughs made with and without preferments.

PREFERMENTS

A *preferment* is a mixture of dough made prior to mixing the final dough. The preferment uses a portion of the total formula's flour, water, yeast, and sometimes salt (see note). This mixture is allowed to ferment for at least 3 hours (and often much longer) before being added to the final bread dough during mixing. The initial fermentation provides the final bread dough with the benefit of a long fermentation without its needing to ferment after the dough is mixed. Additional benefits are numerous, and include the production of essential acidity, alcohol, and gases, as well as flavors and aromas. In addition, using preferments decreases the amount of yeast needed in the final dough formula. It also increases the strength of the final dough and creates a more thoroughly hydrated dough, which extends the shelf life of the bread.

While the advantages of using preferments are numerous, there are a few negatives as well. First and foremost, preferments require extra time and planning. It can also be difficult to predict how much preferment to make, should there be changes in production quantities. In addition, using preferments requires more space in the bakeshop, either in the refrigerator for cold storage or on a counter at room temperature. For large-scale production, this space requirement can be quite a drawback, so it is ideal to have a reserved space for preferments, preferably one that is temperature controlled.

These drawbacks do not outweigh the positives of using preferments in the bakeshop, but bakers should be aware of them as they plan their formulas and production.

Note: The quantities of ingredients used in the preferment can be calculated using the bakers' percentage. Example formulas for these calculations can be found in Chapter 4 (see pages 78–81). Generally, preferments use 10 to 30 percent of the total flour in the formula, though some use up to 60 percent, most commonly for Italian breads. When a preferment is used, the yeast in the final dough should be reduced by 25 to 50 percent.

Mixing the Preferment

As with all baking, precise scaling is the first and most important step. After this basic step, however, mixing the preferment varies only slightly from the standard methods of bread mixing. Importantly, the fermentation needs to be strictly regulated to ensure a consistent end result. In general, the water temperature should be around 60°F/16°C. The temperature can be increased to shorten the pre-fermentation time, but it should not be decreased if a longer pre-fermentation time is desired. Water that is too cool can damage the yeast; to lengthen the time, it is better to just reduce the quantity of yeast in the preferment.

The primary function of a preferment is to bring flavor to the dough, so it is not necessary to develop the gluten structure or encourage gas retention. The mixing needs to be just long enough to fully blend the ingredients without incorporating too much oxygen. A dough hook should be used unless the preferment is extremely liquid, in which case the paddle attachment will more likely incorporate the ingredients quickly. The preferment is then transferred to a storage container and held the required amount of time. It is advised to use a large container to store the preferment, as the space will allow the preferment to increase in volume.

Incorporating Preferments into the Final Dough

Some pre-fermented dough is added half way through the mixing to avoid overmixing that portion of the dough. Other preferments, like a poolish, can be added either at the very beginning of the mixing process or after the pickup stage of mixing. If using a preferment during autolyse (see page 93), the preferment should be added to the dough along with the yeast and salt only after the resting period; this avoids incorporating any yeast during the autolyse stage itself. Sourdough and levain (see pages 145 and 132) are the exceptions to this rule. Because these starters have a slower fermentation, they can be incorporated before the autolyse begins. However, if the water temperature at the time of mixing is very cold, add the preferment after mixing to avoid delaying the fermentation of wild yeast cultures.

Note: Preferments can also be used to adjust the temperature of the water being added to the dough during mixing. For example, if a preferment has been refrigerated, it can be added straight to the dough to eliminate or lessen the need to use cold water or ice in the heat of summer, when room temperatures are extremely high.

Other Effects of Preferments

In addition to the positive and negative impacts of preferments listed earlier, there are a few secondary effects as well.

- When flour and water combine and begin enzyme activity, this activity generates sugar degradation (amylase) and protein degradation (protease).
- Occasionally the interior of a preferment begins to liquefy, especially near the end of maturation. This happens because of excess enzyme activity, but it can be prevented by adding 0.1 to 0.2 percent salt to the preferment.
- During pre-fermentation, the yeast begins to consume the sugars naturally present in the carbohydrates of the flour. This is especially true of long fermentation times at room temperature. When the preferment is added to the final dough, the overall amount of sugars available to encourage fermentation is lower than when a preferment is not used. This can result in less coloration of the crust, but can be combated by adding 0.5 to 1 percent diastatic malt to the final dough.

The baguette on the left was made using a poolish; the center baguette was made with a *pâte fermentée*; the baguette on the right was made with a liquid levain.

- The more liquid preferments, such as poolish (see page 124), promote higher levels of enzyme activity. As a result, the dough will be more extensible, which reduces the mixing time, helps prevent overoxidization, and eases the shaping process. These factors also give the finished loaf more volume and a more open crumb structure.
- The flavor and aroma of preferments are affected by many factors: the level of hydration in the preferment, whether it is fermented at room temperature or in the refrigerator, the addition of salt, the type of yeast, and so on.

Types of Preferments

There are several preferments that can be used in the production of lean doughs. Each type is different, and each has a different impact on the finished dough. Some varieties of preferments include pre-fermented dough, poolish, biga, and levain.

Pre-fermented Dough

Pre-fermented dough is also known as *pâte fermentée* or "pate ferment." The concept was born of necessity and compromise: to compensate for the poor-quality dough being produced in bakeries that didn't have the time or equipment to allow for a long first fermentation. Essentially, a piece of dough is allowed to ferment prior to the final mixing of the dough. This piece could be a portion of dough from a previous mixing that had been saved or be a separate mix done for the sole purpose of making a preferment. Different types of preferments yield different results (see photo on page 122). The crust of a dough made with *pâte fermentée* is not as flaky as one made with poolish.

Ratio: 100% flour, 60 to 65% water, 0.5 to 1% yeast, 2.2% salt

Fermentation Time: 3 to 6 hours at room temperature prior to mixing the final dough. If production schedules require that the preferment be held longer (or if the baker is using a portion of dough from the previous day's mixing), it should be fermented for 1 to 2 hours at room temperature, then held under refrigeration until the final dough is mixed (up to 24 hours). If preferment is refrigerated, it should be brought to room temperature 1 to 2 hours before the final mixing so as to not adversely affect the temperature of the dough; otherwise, the water temperature will need to be adjusted to control the final desired dough temperature (DDT; see page 82).

Use: Easily workable into the dough and can be used in a variety of products, from breads and rolls to yeast-raised pastries. Needs to be held under refrigeration if it isn't used within 6 hours (which can take up a large quantity of refrigerator space).

POOLISH

Poolish was created by Polish bakers near the end of the nineteenth century. Using commercial yeast and high levels of hydration to create a loose preferment that was similar in consistency to liquid levain (see pages 132–133), bakers discovered that the resulting bread was lighter and less acidic than sourdough bread. This flavor became popular, and thus the use of poolish spread into Austria and France. The crust of a bread made with poolish also has a thinner, flakier crust than one made with liquid levain.

Ratio: 100% flour, 100% water, 0.1 to 0.5% yeast. Note: Because poolish ferments at room temperature, the levels of yeast vary based on fermentation time. The longer the fermentation time before the poolish is added to the dough, the less yeast should be used. Likewise, water temperatures may need to be adjusted to compensate for excessively warm temperatures in the bakeshop to ensure the poolish does not overferment.

Fermentation: Poolish should be fermented for 16 to 18 hours at room temperature, and bakers should time the poolish to be fully matured at the precise time the final dough is ready to be mixed. A fully mature poolish will have foamed at the top and then begun to recede. The longer a poolish ferments, the more aroma and flavor it develops and the less yeast the final dough requires.

Use: Very liquid when fully matured; therefore, it is easier to measure out into quantities for various doughs just after mixing it than to divide it once it has fermented.

A poolish before and after fermentation.

Poolish mixes very easily into final doughs and is generally added at the beginning of the mixing, owing to its high water content. Poolish is most commonly used for the production of baguettes.

BIGA

Biga is an Italian preferment that traditionally was stiff, used to add and/or reinforce the strength of the finished dough. Through the years, formulas that included biga yielded varying results—some were stiff, some were very liquid, and others were sour. Research finally concluded that *biga* was simply the Italian word for "preferment," and could mean any type of preferment to be used in an Italian bread recipe. This book uses the traditional definition of the term, producing a stiff preferment that yields a strong end result. Bakers should be aware, however, that owing to the increased strength of flour, bigas need to be used properly to prevent a negative impact on extensibility.

Ratio: 100% flour, 50 to 60% water, 0.1% to 0.5% yeast

Fermentation Time: Biga ferments at room temperature for 16 to 18 hours, based on a relatively neutral room temperature of 60°F/15°C. For higher room temperatures, the water temperature, the amount of yeast, and possibly the fermentation time should be adjusted to compensate and prevent overfermentation.

A biga before and after fermentation.

Use: More difficult to add to a dough because of its firmer, stiffer texture; care should be taken to ensure proper mixing. Biga increases the strength of doughs, which makes it ideal for doughs with high hydration, such as ciabatta or focaccia.

SOURDOUGH STARTER

A sourdough starter before and after fermentation.

The process of making sourdough bread begins with creation of a *sour*, a culture of microorganisms that are fed and cultivated to increase their quantity. This sour is then used to ferment the final bread dough. To fully understand the process of sourdough baking, it is important to examine each stage, step by step.

A sourdough is commonly referred to as a *levain*. This is simply the French term for the preferment, and the terms can be used interchangeably. A liquid levain (see pages 132–133) is a common variation of a sourdough with an increased hydration level.

MICROORGANISMS AT WORK There are two primary microorganisms that help create the sourdough culture: yeast and bacteria. Both need different things to thrive and therefore are affected differently by every factor in the process, including hydration levels, other ingredients, temperatures, and acidity.

The yeasts present in a sour culture are classified as "wild yeast" because they are naturally present in the environment. The most predominant yeast in sour cultures is *Saccharomyces cerevisiae*. This yeast consumes the simple sugars (glucose and fructose), producing alcohol (ethanol) and gas (carbon dioxide) as the dough begins to ferment. Just as with commercial yeast, the gas contributes to the development of the dough, and the alcohol contributes flavors and aromas.

Liquid levain before and after fermentation.

The bacteria are lactic bacteria (*Lactobacillus* and/or *Lactococcus*). These bacteria convert sugars into organic acids, which produce additional flavors and aromas in the bread. Again, the results are similar to those produced with commercial yeast.

These microorganisms are everywhere, but they are specifically prominent in the flour being used to create the sour starter. One gram of flour contains roughly 13,000 wild yeast cells and 320 lactic bacteria cells. It is ideal to use organic flours, which produce the best sour development.

The sourdough process can be understood as six steps:

1. Create a culture.
2. Elaborate the culture.
3. Prepare the culture for use (now referred to as a "starter").
4. Feed the starter (after one or more feedings, the culture is referred to as a "feeding"; after the final feeding, the culture is called a "levain").
5. Perpetuate the culture.
6. Incorporate the levain into the final dough.

STARTING A SOUR CULTURE There are many ways to start a sourdough culture, but the concept is always the same. The initial microorganisms are present in the flour, but to develop them they need to be activated enough to begin to ferment the dough. This happens when the microorganisms are fed, causing them to reproduce and begin to produce alcohol, gas, and acidity. This food comes from additions of flour, water, and oxygen.

Some formulas call for the addition of simple sugars in the form of malt or honey, which also provide immediate food and nutrients for the microorganisms.

Note: Using organic flour can increase the chances of creating a successful starter. Because no chemicals or pesticides are used on the wheat ground into the flour, organic flours contain higher levels of the naturally present microorganisms. Rye flour is also richer in these microorganisms along with a variety of minerals. The minerals can serve as additional food for the yeast and bacteria.

CULTURE ELABORATION Culture elaboration is the process that occurs from the creation of the culture until it is used as a starter. When the ingredients are mixed, the hydration of the flour and the incorporation of oxygen start the activity of the microorganisms. Initially, sufficient oxygen introduction creates the ideal conditions for reproduction of the microorganisms. After a few hours, the increase in microorganisms begins to reduce the amount of available oxygen, eventually creating an anaerobic environment. At this point, fermentation begins, enhanced by consistently warm temperatures. After about 22 hours, the culture will have doubled in volume.

During elaboration, there's a balance of yeast and bacteria, in both quantity and quality. Some of the microorganisms are more or less resistant to the lack of food, lack of oxygen, or acidification of the culture. Cohabitation of the yeast and bacteria is also possible because they are not competing for the same nutrients. But to ensure that the yeast and bacteria can continue to reproduce and begin fermentation, the vital conditions (food, water, and oxygen) of the culture need to be renewed. This happens through the course of feeding the starter, which is several times during elaboration.

While a schedule is to be followed, a baker can tell that a culture needs a feeding when it begins to collapse in the center. A well-established culture should rise four times in volume during the 6 to 8 hours of fermentation. When this volume is reached, the culture has become a starter. It is important to remember that even though the primary type of yeast and bacteria are the same, each culture will be different and will produce different products.

CREATING A SOUR STARTER Sour cultures need regular feedings for the first five days to become useable sour starters. During this process, the culture should be held at 80°F/27°C to encourage proper fermentation. While there are a variety of recipes for sourdough breads in this book, the following is an example of a starter feeding schedule so as to get a sense of timing and ratios. After five to eight days of feeding, the starter may be changed into a sour. A good indicator of this is when the starter changes from a dome shape to a more concave shape.

FEEDING THE STARTER The schedule for the starter can be adjusted to meet the needs of the bakeshop once it is fully developed. Follow the schedule below to create a fully developed starter that can be used after five days. At that point, the sourdough starter can be refrigerated to slow its fermentation and space out the feeding schedule. The starter can also be dried out with flour and frozen to preserve it for further use.

SCHEDULE	FLOUR	WATER	STARTER	NEXT FEEDING
Day 1 (AM)	1 lb 1¾ oz/500 g whole wheat flour 1 lb 1¾ oz/500 g bread flour (½ oz/15 g malt; optional)	2 lb 3¼ oz/1 kg		24 hours
Day 2 (AM)	1 lb 1¾ oz/500 g bread flour	1 lb 1¾ oz/500 g	1 lb 1¾ oz/500 g	6 to 8 hours
Day 2 (PM)	1 lb 1¾ oz/500 g bread flour	1 lb 1¾ oz/500 g	1 lb 1¾ oz/500 g	16 hours
Day 3 (AM)	1 lb 1¾ oz/500 g bread flour	1 lb 1¾ oz/500 g	1 lb 1¾ oz/500 g	6 to 8 hours
Day 3 (PM)	1 lb 1¾ oz/500 g bread flour	1 lb 1¾ oz/500 g bread flour	1 lb 1¾ oz/500 g bread flour	16 hours
Day 4 (AM)	1 lb 1¾ oz/500 g bread flour	1 lb 1¾ oz/500 g bread flour	1 lb 1¾ oz/500 g bread flour	6 to 8 hours
Day 4 (PM)	1 lb 1¾ oz/500 g bread flour	1 lb 1¾ oz/500 g bread flour	1 lb 1¾ oz/500 g bread flour	16 hours
Day 5 (AM)	1 lb 1¾ oz/500 g bread flour	1 lb 1¾ oz/500 g bread flour	1 lb 1¾ oz/500 g bread flour	6 to 8 hours
Day 5 (PM)	1 lb 1¾ oz/500 g bread flour	1 lb 1¾ oz/500 g bread flour	1 lb 1¾ oz/500 g bread flour	16 hours

Several factors can affect the activity of the culture during the feeding process. This, in turn, affects the characteristics of the finished loaf.

- **Hydration:** A firm or stiff culture will more likely develop acetic acidity, while more liquid cultures more likely will develop lactic acidity. Water with high chlorine content can delay the activity of the culture.
- **Flour:** Activity of the enzymes naturally present in the flour will determine the amount of sugars available for consumption by the microorganisms. Bran, which is rich in mineral content, will provide additional food for the yeast and bacteria. Flour with higher extraction provides a better basis for activity and acidity production. Rye flour is especially high in nutrients and can assist in all levels of microbiological activity, but it can reduce the final volume of the loaf.

- **Temperature:** Warmer temperatures favor bacterial activity and the production of lactic acidity. However, high temperatures also increase yeast activity, making fermentation difficult to control. Optimal temperatures are around 77°F/25°C. When temperatures are too low, fermentation is suppressed and production of acetic acid rises.
- **Salt:** Small quantities of salt can be beneficial for cultures with excessive extensibility and high protease activity (up to 0.1%). Amounts higher than 0.1 percent can inhibit the activity of some types of yeast and bacteria by delaying or stopping fermentation.

MAINTAINING THE CULTURE To maintain a pure, consistent culture, bakers should take care to properly sanitize all work areas and equipment prior to feeding the starter. The starter should not be contaminated with scraps of other dough, especially dough being

Sourdough Troubleshooting

PROBLEM(S)	SOURCE	SOLUTIONS
Lack of acidity Lack of strength Poor flavor in finished loaf	Sour is too young Lack of fermentation activity Not allowed to ferment long enough between feedings Not enough sour used in finished dough	Increase maturation time for the sour Increase room temperature for fermentation times Use flour with higher quantities of bran Ensure water is not heavily chlorinated Increase fermentation time for the bread dough
Excessive acidity Sharp flavor	Dough liquification Old sour Feeding schedule inconsistent Overly lengthy fermentation between feedings Bacterial activity too high Too much sour used in final dough	Start a new culture Shorten fermentation time between feedings Add salt to decrease activity levels Use less sour in final dough
Lack of bread development	Not enough fermentation activity Not enough yeast production Excessive acidity (inhibiting yeast activity) Stored at too cool a temperature	Add a small amount (up to 2% of commercial yeast) Make final dough warmer and softer Keep sourdough above 50°F/10°C
Lack of strength	Lack of acidity Lack of gas production Dough too cold and the end of mixing Not enough sour used in final dough	Longer first fermentation More folds Increase fermentation temperature Increase amount of sour used in final dough

made with commercial yeast. The feeding process should remain consistent—maintaining the proper proportion of ingredients, the precise feeding schedule, correct temperatures, and exact fermentation times.

GRAIN AND SEED SOAKERS

Whole grains and seeds make excellent additions to breads, adding flavor, texture, and nutritional benefits. Before being added to the bread dough, however, the grains (and some seeds) should be soaked for several reasons. For one, most grains and seeds are very hard, and they will not fully soften when baked into a loaf, negatively affecting the texture of the finished bread. Whole wheat or rye berries must be soaked overnight to soften them to a point where they are edible.

In addition, grains and seeds can absorb significant quantities of moisture; if they are not soaked before being mixed into a bread dough, they will absorb that moisture from the dough itself. This can adversely affect the dough, altering the hydration of the flour and damaging the developing gluten bonds.

Soakers can be made in two different ways: cold soaking and hot soaking.

Cold Soakers

Cold soakers must be made at least one day in advance. The grains and liquid are incorporated, covered, and allowed to soak overnight. The cold soaking method prevents the starches from becoming gelatinized while softening the grains, making them easier to add to bread recipes.

Hot Soakers

Hot soakers pre-gelatinize the starches of the grains. This method is faster and reduces the baking time of some breads, but bakers often feel there is a loss of flavor and quality of texture. Hot soakers are made by bringing the liquid to a boil, then adding the grains. The mixture is then cooked for 5 minutes over low heat, and set aside to cool for a minimum of 1 hour prior to being added to the dough. Hot soakers are necessary for certain specialty breads, such as stollen, and when using whole grains, such as rye or wheat berries.

Adding the Soaker to the Dough

Soakers should be added after the dough has begun to develop, generally when the mixer speed has been raised to medium for the intermediate stage of mixing. The soaker is added and mixed on medium speed for several minutes, until it is fully and evenly incorporated. The exception to this rule is rye bread. Owing to the lower gluten development in a rye bread, soakers should be added to rye doughs at the beginning of the mixing time.

Liquid Levain

FIRST FEEDING

FERMENTATION TIME: 12 HOURS

WATER TEMPERATURE: 80°F/27°C

INGREDIENT	US	METRIC	BAKERS' PERCENTAGE
Water	4½ oz	125 g	125%
Honey	Trace amount		
Organic medium coarse rye flour	3½ oz	100 g	100%
Total	**8 oz**	**125 g**	**225%**

Methods:

1. Mix water, honey, and rye flour together until homogenous.
2. Cover and allow to sit at room temperature for 12 hours.

SECOND FEEDING–EIGHTH FEEDING

FERMENTATION TIME: 12 HOURS

WATER TEMPERATURE: 80°F/27°C

INGREDIENT	US	METRIC	BAKERS' PERCENTAGE
Water	4½ oz	125 g	125%
Mix previous feeding	4 oz	115 g	115%
Organic medium coarse rye flour	3½ oz	100 g	100%
Total	**12 oz**	**342 g**	**340%**

Methods:

1. Measure 115 g of mixture from previous feeding and discard excess.
2. Add water to mixture to dissolve. Add flour and mix until homogenous.
3. Cover and allow to sit at room temperature for 12 hours.

NINTH FEEDING AND BEYOND

FERMENTATION TIME: 12 HOURS

WATER TEMPERATURE: 80°F/27°C

INGREDIENT	US	METRIC	BAKERS' PERCENTAGE
Water	10³⁄₅ oz	300 g	125%
Mix previous feeding	1¾ oz	50 g	20.8%
Organic medium coarse rye flour	8½ oz	240 g	100%
Total	**1 lb 4⁵⁄₆ oz**	**590 g**	**245.8%**

Methods:

1. Measure 50 g of mixture from previous feeding and discard excess.
2. Add water to mixture to dissolve. Add flour and mix until homogenous.
3. Cover and allow to sit at room temperature for 12 hours.

CHEF'S NOTES

After using liquid levain, save 50 g to continue feeding for future baking.

To preserve your liquid levain without daily feeding: Feed as detailed in the ninth feeding above and then place it immediately in the refrigerator. The liquid levain can stay in the refrigerator for several weeks. Note that some gray water may form on top of the levain when refrigerated. The color is due to the natural color of the flour, and the water should be stirred back into the levain.

To refresh the refrigerated liquid levain before use: Remove from refrigerator 3 days in advance, feeding every day as detailed in the ninth feeding above.

Baguette

DESIRED DOUGH TEMPERATURE: 78°F/26°C

INGREDIENT	US	METRIC	BAKERS' PERCENTAGE
Water	6 lb	2.7 kg	69.0%
Malt syrup	1¼ oz	35 g	0.9%
Bread flour	8 lb 9½ oz	3.9 kg	100.0%
Yeast, instant dry	1 oz	30 g	0.7%
Salt	3¼ oz	90 g	2.3%
Total Dough Weight	15 lb	6.8 kg	172.9%

1. Combine the water, malt syrup, flour, yeast, and salt in a mixer and mix on low speed until homogenous, about 4 minutes. Increase the speed to high and mix until the dough reaches the improved stage of gluten development, about 3 minutes.
2. Bulk-ferment, covered, for about 2 hours, folding the dough once halfway through fermentation.
3. Divide the dough into 13½ oz/380 g pieces and preshape into logs. Bench-rest the dough, covered, for approximately 15 minutes.
4. Shape the logs into 21-in/53-cm baguettes and place them seam side down on a couche-covered wooden board. There should be room for 8 baguettes per board. Proof the dough for about 20 minutes or as needed, depending on the environment.
5. Score the baguettes 5 times (see page 107) and load them into a 470°F/243°C deck oven. Steam 1 second before loading and 3 seconds after loading.
6. Bake the baquettes until deep golden brown, about 25 minutes. Open the vent for the last 6 minutes of baking time to develop the crust.
7. Place the finished breads on a rack to cool.

CHEF'S NOTE

This dough can be used for a number of French regional items, such as French rolls, Pan Tricorne (page 11), and Pain des Venganges.

To make Epi Baguette, place a proofed baguette on the loader. Dust the baguette with bread flour, if desired. Place scissors flat against the top of the dough and make a cut that goes almost through to the bottom of the dough, being sure not to cut all the way through. Turn the dough piece to one side. Make another cut 2 in/5 cm from the first cut and turn the dough piece to the opposite side. Repeat the cuts and angling of the dough pieces to the end of the baguette. As you get closer to the end of the baguette, make the cuts a little closer together. Be sure that the baguette is not overproofed or it will collapse when it is cut. The finished loaf should look like a stalk of wheat.

Baguette with Liquid Levain

DESIRED DOUGH TEMPERATURE: 78°F/26°C

INGREDIENT	US	METRIC	BAKERS' PERCENTAGE
Water	6 lb 2¾ oz	2.8 kg	62.0%
Bread flour	11 lb	5 kg	100.0%
Liquid Levain (page 132)	1 lb 4¾ oz	590 g	11.2%
Yeast, instant dry	4 oz	115 g	2.0%
Sea salt	2 oz	55 g	1.0%
Total Dough Weight	18 lb 13½ oz	8.6 kg	176.2%

1. To prepare the final dough, combine the water, flour, and liquid levain in a mixer. Mix on low speed until homogenous, about 1 minute. Sprinkle the yeast on top of the dough and let rest in the mixer for about 15 minutes to autolyse.
2. Resume mixing on low speed for an additional minute. Add the salt and mix for 30 seconds. Increase the speed to high and mix until the dough reaches the improved stage of gluten development, about 2 minutes.
3. Bulk-ferment, covered, for about 2 hours, folding the dough once halfway through fermentation.
4. Divide the dough in 13½ oz/380 g pieces and preshape into logs. Bench-rest the dough, covered, for approximately 15 minutes.
5. Shape the logs into 21-in/53-cm baguettes and place them seam side down on a couche-covered wooden board. There should be room for 8 baguettes per board. Proof the dough for about 20 minutes or as needed, depending on the environment.
6. Score the baguettes 5 times (see page 107) and load them into a 470°F/243°C deck oven. Steam 1 second before loading and 3 seconds after loading.
7. Bake until deep golden brown, about 25 minutes. Open the vent for the last 6 minutes of baking time to develop the crust.
8. Place the finished breads on a rack to cool.

Baguette with Pâte Fermentée

DESIRED DOUGH TEMPERATURE: 78°F/26°C

INGREDIENT	US	METRIC	BAKERS' PERCENTAGE
Water	7 lb 11½ oz	3.5 kg	70.2%
Bread flour	11 lb	5 kg	100.0%
Pâte Fermentée (page 199)	2 lb 3 oz	990 g	19.7%
Malt syrup	1 oz	40 g	0.8%
Yeast, instant dry	¾ oz	20 g	0.4%
Salt	4 oz	115 g	2.3%
Total Dough Weight	21 lb 4¼ oz	9.7 kg	193.4%

1. To prepare the final dough, combine the water, flour, pâte fermentée, and malt syrup in a mixer. Mix on low speed until homogenous, about 1 minute. Sprinkle the yeast on top of the dough and let rest in the mixer for about 15 minutes to autolyse.
2. Resume mixing on low speed for an additional minute. Add the salt and mix for 30 seconds. Increase the speed to high and mix until the dough reaches the improved stage of gluten development, about 2 minutes.
3. Bulk-ferment, covered, for about 2 hours, folding the dough once halfway through fermentation.
4. Divide the dough in 13½ oz/380 g pieces and preshape into logs. Bench-rest the dough, covered, for approximately 15 minutes.
5. Shape the logs into 21-in/53-cm baguettes and place them seam side down on a couche-covered wooden board. There should be room for 8 baguettes per board. Proof the dough for about 20 minutes or as needed, depending on the environment.
6. Score the baguettes 5 times (see page 107) and load them into a 470°F/243°C deck oven. Steam 1 second before loading and 3 seconds after loading.
7. Bake until deep golden brown, about 25 minutes. Open the vent for the last 6 minutes of baking time to develop the crust.
8. Place the finished breads on a rack to cool.

Baguette with Poolish

DESIRED DOUGH TEMPERATURE: 78°F/26°C

INGREDIENT	US	METRIC	BAKERS' PERCENTAGE
Poolish			
Water	2 lb 3¼ oz	1 kg	100.0%
Bread flour	2 lb 3¼ oz	1 kg	100.0%
Yeast, instant dry	¼ oz	5 g	0.1%
Final Dough			
Water	5 lb 1 oz	2.3 kg	56.3%
Bread flour	11 lb	5 kg	100.0%
Poolish	4 lb 6½ oz	2 kg	50.0%
Yeast, instant dry	1½ oz	40 g	0.4%
Sea salt	3½ oz	100 g	2.5%
Total Dough Weight	18 lb 8 oz	8.4 kg	209.2%

1. Prepare the poolish about 18 hours before mixing the final dough.
2. To prepare the final dough, combine the poolish, water, and flour in a mixer. Mix on low speed until homogenous, about 1 minute. Sprinkle the yeast on top of the dough and let rest in the mixer for about 15 minutes to autolyse.
3. Resume mixing on low speed for an additional minute. Add the salt and mix for 30 seconds. Increase the speed to high and mix until the dough reaches the improved stage of gluten development, about 2 minutes.
4. Bulk-ferment, covered, for about 2 hours, folding the dough once halfway through fermentation.
5. Divide the dough in 13½ oz/380 g pieces and pre-shape into logs. Bench-rest, covered, for approximately 15 minutes, depending on the temperature and tightness of the dough.
6. Shape the logs into 21-in/53-cm baguettes and place them seam side down on a couche-covered wooden board. There should be room for 8 baguettes per board. Proof the dough for about 20 minutes or as needed, depending on the environment.
7. Score the baguettes 5 times (see page 107) and load them into a 470°F/243°C deck oven. Steam 1 second before loading and 3 seconds after loading.
8. Bake until deep golden brown, about 25 minutes. Open the vent for the last 6 minutes of baking time to develop the crust.
9. Place the finished breads on a rack to cool.

Couronne Bordelaise

DESIRED DOUGH TEMPERATURE: 78°F/26°C

INGREDIENT	US	METRIC	BAKERS' PERCENTAGE
Water	6 lb	2.7 kg	69.0%
Malt syrup	1¼ oz	35 g	0.9%
Bread flour	8 lb 9½ oz	3.9 kg	100.0%
Yeast, instant dry	1 oz	30 g	0.7%
Salt	3¼ oz	90 g	2.3%
Total Dough Weight	14 lb 15 oz	6.8 kg	172.9%
Vegetable oil, for brushing	As needed	As needed	

1. Combine the water, malt syrup, flour, yeast, and salt in a mixer and mix on low speed until homogenous, about 4 minutes. Increase the speed to high and mix until the dough reaches the improved stage of gluten development, about 3 minutes.
2. Bulk-ferment, covered, for about 1 hour, folding the dough once halfway through fermentation.
3. Divide the dough in 4 lb/1.8 kg presses and preshape into rounds. Bench-rest the dough, covered, for approximately 15 minutes.
4. Use a dough divider to divide the presses into 18 pieces. For each loaf, you will need 10 pieces of dough: 9 single rolls and 1 double roll (two 3½ oz/100g pieces rolled together). Shape each piece into rounds.
5. Use a rolling pin to roll the larger piece into a 11-in/28-cm circle. Lay the circle over the center hump of a lined couronne banneton, and press it into the basket. Brush the outside edge with oil.
6. Place the remaining rounds seam side up into the outside of the basket, covering the edge of the large circle. They will fit tightly.
7. Use a paring knife to cut the thin dough on the center hump of the banneton. Cut the dough into 9 even triangles so that each triangle lines up with one of the dough rounds. Fold the triangles over the rounds.
8. Proof the dough for about 20 minutes, or as needed, depending on the environment.
9. Load the dough into a 500°F/260°C deck oven. Immediately lower the temperature to 460°F/238°C and bake until golden brown, about 18 minutes. Open the vent and bake until dark brown, especially around the edges, about 6 minutes.
10. Place the finished breads on a rack to cool.

Filone

DESIRED DOUGH TEMPERATURE: 80°F/27°C

INGREDIENT	US	METRIC	BAKERS' PERCENTAGE
Biga, 85% hydration (page 125)	5 lb 8 oz	2.5 kg	50.0%
Water	6 lb 13 oz	3.1 kg	62.0%
Bread flour	11 lb	5.0 kg	100.0%
Yeast, instant dry	2¾ oz	75 g	1.5%
Sea salt	4½ oz	125 g	2.5%
Total Dough Weight	23 lb 12½ oz	10.8 kg	216.0%

1. Prepare the biga the day before mixing the final dough.
2. Combine the biga, water, and flour in a mixer and mix until homogenous, about 1½ minutes. Stop the mixer, sprinkle the yeast on top of the dough, and rest to autolyse for 10 to 15 minutes.
3. Resume mixing to incorporate the yeast. Add the salt and mix until combined, about 30 seconds. Increase the speed to high and mix until the dough reaches the improved stage of gluten development, about 2 minutes.
4. Remove the dough from the mixer and bulk-ferment, covered, until doubled in size, about 1½ hours. Fold once halfway through fermentation.
5. Divide the dough into 1 lb/450 g pieces and pre-shape into rounds. Bench-rest the dough for about 15 minutes.
6. Shape the dough into filones by folding the top and bottom edges toward the center of the dough, sealing the seam with the palm of your hand. Shape into a 12-in/30-cm log. Place seam side down on a well-floured couche.
7. Proof the dough for about 1 hour or as needed, depending on the environment.
8. Load the dough seam side up into a 425°F/218°C oven (the seams will burst open during baking). Steam for 1 second before loading and 3 seconds after loading. Bake until dark golden brown, about 15 minutes.
9. Place the finished breads on a rack to cool.

White Sourdough Starter

FIRST FEEDING

FERMENTATION TIME: 12 HOURS

WATER TEMPERATURE: 80°F/27°C

INGREDIENT	US	METRIC	BAKERS' PERCENTAGE
Water	4 oz	114 g	100%
Organic bread flour	4 oz	114 g	100%
Total	**8 oz**	**229 g**	**200%**

Methods:

1. Mix water and bread flour together until homogenous.
2. Cover and allow to sit at room temperature for 12 hours.

SECOND FEEDING

FERMENTATION TIME: 12 HOURS

Methods:

1. Recombine mixture from day one.
2. Cover and allow to sit at room temperature for 12 hours.

THIRD FEEDING

FERMENTATION TIME: 12 HOURS

WATER TEMPERATURE: 80°F/27°C

INGREDIENT	US	METRIC	BAKERS' PERCENTAGE
Water	4 oz	114 g	100%
Mix feeding 1, 2	4 oz	114 g	100%
Organic bread flour	4 oz	114 g	100%
Total	**12 oz**	**342 g**	**300%**

Methods:

1. Measure 114 g of mixture from first and second feedings and discard excess.
2. Add water to mixture to dissolve. Add flour and mix until homogenous.
3. Cover and allow to sit at room temperature for 12 hours.

FOURTH FEEDING

FERMENTATION TIME: 12 HOURS

WATER TEMPERATURE: 80°F/27°C

INGREDIENT	US	METRIC	BAKERS' PERCENTAGE
Water	4 oz	114 g	100%
Mix from third feeding	8 oz	227 g	200%
Organic bread flour	4 oz	114 g	100%
Total	**1 lb**	**455 g**	**400%**

Methods:

1. Measure 227 g of mixture from third feeding and discard excess.
2. Add water to mixture to dissolve. Add flour and mix until homogenous.
3. Cover and allow to sit at room temperature for 12 hours.

FIFTH FEEDING

FERMENTATION TIME: 12 HOURS

WATER TEMPERATURE: 80°F/27°C

INGREDIENT	US	METRIC	BAKERS' PERCENTAGE
Water	8 oz	227 g	66.7%
Mix from fourth feeding	4 oz	113 g	33.3%
Organic bread flour	12 oz	340 g	100.0%
Total	**1 lb 8 oz**	**680 g**	**200%**

Methods:

1. Measure 113 g of mixture from third feeding and discard excess.
2. Add water to mixture to dissolve. Add flour and mix until homogenous.
3. Cover and allow to sit at room temperature for 12 hours.

SIXTH FEEDING

FERMENTATION TIME: 12 HOURS

WATER TEMPERATURE: 80°F/27°C

INGREDIENT	US	METRIC	BAKERS' PERCENTAGE
Water	8 oz	227 g	66.7%
Mix from fifth feeding	4 oz	113 g	33.3%
Organic bread flour	12 oz	340 g	100.0%
Total	**1 lb 8 oz**	**680 g**	**200%**

Method (see fifth feeding)

SEVENTH FEEDING

FERMENTATION TIME: 12 HOURS

WATER TEMPERATURE: 80°F/27°C

INGREDIENT	US	METRIC	BAKERS' PERCENTAGE
Water	8 oz	227 g	66.7%
Mix from sixth feeding	4 oz	113 g	33.3%
Organic bread flour	12 oz	340 g	100.0%
Total	**1 lb 8 oz**	**680 g**	**200%**

Method (see fifth feeding)

EIGHTH FEEDING

FERMENTATION TIME: 12 HOURS

WATER TEMPERATURE: 80°F/27°C

INGREDIENT	US	METRIC	BAKERS' PERCENTAGE
Water	8 oz	227 g	66.7%
Mix from seventh feeding	4 oz	113 g	33.3%
Organic bread flour	12 oz	340 g	100.0%
Total	**1 lb 8 oz**	**680 g**	**200%**

Method (see fifth feeding)

CHEF'S NOTE

To keep this white sourdough starter alive, continue feeding as detailed in the fifth feeding above.

To preserve your white sourdough starter without daily feeding: Feed as detailed in the fifth feeding above and then place it immediately in the refrigerator. The white sourdough starter can stay in the refrigerator for 3 weeks.

To refresh the refrigerated white sourdough starter before use: Remove from refrigerator 3 days in advance, feeding every day as detailed in the fifth feeding above.

Sourdough Bread

DESIRED DOUGH TEMPERATURE: 76°F/24°C

INGREDIENT	US	METRIC	BAKERS' PERCENTAGE
White Sourdough (page 142)	4 lb 6½ oz	2 kg	39.3%
Water	7 lb 11 oz	3.5 kg	69.7%
Malt syrup	1½ oz	40 g	0.8%
Bread flour	10 lb 5¾ oz	4.7 kg	93.2%
Whole wheat flour	12 oz	340 g	6.8%
Salt	4¾ oz	135 g	2.7%
Total Dough Weight	23 lb 9½ oz	10.7 kg	212.5%

1. Prepare the final dough a day ahead. Combine the white sourdough, water, malt syrup, bread flour, and whole wheat flour in a mixer. Mix on low speed until the dough is homogenous, about 4 minutes. Allow the dough to rest in the mixer for 15 to 20 minutes.
2. Mix the dough on low speed for an additional 2 minutes, adding the salt during the last 30 seconds of low speed. Increase the speed to high and mix until the dough reaches the improved stage of gluten development, about 2 minutes.
3. Remove the dough from the mixer and bulk-ferment, covered, for 2 hours. Fold once halfway through fermentation.
4. Divide the dough into 1 lb 8 oz/680 g pieces and preshape into rounds. Bench-rest the dough for about 15 minutes.
5. Shape the dough into boules or bâtards and place in lightly floured bannetons. Proof for about 45 minutes or as needed, depending on the environment. Place the dough in a refrigerator to retard overnight.
6. Remove the dough from the cooler 1 hour before baking.
7. Score the dough and load into a 490°F/254°C oven. Immediately lower the temperature to 460°F/238°C. Bake until golden, about 18 minutes, followed by an additional 15 minutes with the vent open.
8. Place the finished breads on a rack to cool.

Buttermilk Sourdough Bread

DESIRED DOUGH TEMPERATURE: 76°F/24°C

INGREDIENT	US	METRIC	BAKERS' PERCENTAGE
Water	1 lb 14 oz	860 g	17.2%
Bread flour	10 lb 5¾ oz	4.7 kg	93.3%
Whole wheat flour	12 oz	340 g	6.7%
Malt syrup	1 oz	30 g	0.5%
White Sourdough (page 142)	4 lb 6½ oz	2 kg	39.3%
Buttermilk	4 lb 14½ oz	2.2 kg	44.2%
Milk	1 lb 3½ oz	550 g	11.0%
Salt	5 oz	140 g	2.8%
Total Dough Weight	23 lb 14¼ oz	10.8 kg	215.0%

1. Make the dough a day ahead. Combine the water, bread flour, whole wheat flour, malt syrup, white sourdough, buttermilk, and milk in a mixer. Mix on low speed until homogenous, about 3 minutes. Rest the dough in the mixer for 15 to 20 minutes.
2. Add the salt and mix on low speed until the dough reaches the improved stage of gluten development, about 5 minutes.
3. Remove the dough from the mixer and bulk-ferment for 2 hours. Fold once halfway through fermentation. Fold the dough a final time and ferment for an additional 15 minutes.
4. Divide the dough into 1 lb 8 oz/680 g pieces and preshape into rounds. Bench-rest the dough for 10 to 15 minutes.
5. Shape the dough into boules and place seam side up in lightly floured bannetons. Proof the dough at room temperature for about 1 hour or as needed, depending on the environment. Place in the refrigerator to retard overnight.
6. Remove the dough from the refrigerator 1 hour before baking.
7. Score and load the dough into a 475°F/246°C oven and immediately lower the temperature to 465°F/241°C. Bake for 18 minutes. Open the vent and bake until golden brown, for an additional 15 minutes.
8. Place the finished breads on a rack to cool.

Rye Sourdough Starter

FIRST FEEDING
FERMENTATION TIME: 12 HOURS
WATER TEMPERATURE: 80°F/27°C

INGREDIENT	US	METRIC	BAKERS' PERCENTAGE
Water	4 oz	114 g	100%
Organic medium coarse rye flour	4 oz	114 g	100%
Total	**8 oz**	**229 g**	**200%**

Methods:

1. Mix water and rye flour together until homogenous.
2. Cover and allow to sit at room temperature for 12 hours.

SECOND FEEDING
FERMENTATION TIME: 12 HOURS

Methods:

1. Recombine mixture from day one.
2. Cover and allow to sit at room temperature for 12 hours.

THIRD FEEDING
FERMENTATION TIME: 12 HOURS
WATER TEMPERATURE: 80°F/27°C

INGREDIENT	US	METRIC	BAKERS' PERCENTAGE
Water	4 oz	114 g	100%
Mix feeding 1, 2	4 oz	114 g	100%
Organic medium coarse rye flour	4 oz	114 g	100%
Total	**12 oz**	**342 g**	**300%**

Methods:

1. Measure 114 g of mixture from first and second feedings and discard excess.
2. Add water to mixture to dissolve. Add flour and mix until homogenous.
3. Cover and allow to sit at room temperature for 12 hours.

FOURTH FEEDING

FERMENTATION TIME: 12 HOURS

WATER TEMPERATURE: 80°F/27°C

INGREDIENT	US	METRIC	BAKERS' PERCENTAGE
Water	4 oz	114 g	100%
Mix from third feeding	8 oz	227 g	200%
Organic medium coarse rye flour	4 oz	114 g	100%
Total	**1 lb**	**455 g**	**400%**

Methods:

1. Measure 227 g of mixture from third feeding and discard excess.
2. Add water to mixture to dissolve. Add flour and mix until homogenous.
3. Cover and allow to sit at room temperature for 12 hours.

FIFTH FEEDING–EIGHTH FEEDING

FERMENTATION TIME: 12 HOURS

WATER TEMPERATURE: 80°F/27°C

INGREDIENT	US	METRIC	BAKERS' PERCENTAGE
Water	8 oz	227 g	66.7%
Mix from fourth feeding	4 oz	113 g	33.3%
Organic medium coarse rye flour	12 oz	340 g	100.0%
Total	**1 lb 8 oz**	**680 g**	**200%**

Methods:

1. Measure 113 g of mixture from third feeding and discard excess.
2. Add water to mixture to dissolve. Add flour and mix until homogenous.
3. Cover and allow to sit at room temperature for 12 hours.

CHEF'S NOTE

To keep this rye sourdough starter alive, continue feeding as detailed in the fifth feeding above.

To preserve your rye sourdough starter without daily feeding: Feed as detailed in the fifth feeding above and then place it immediately in the refrigerator. The rye sourdough starter can stay in the refrigerator for 3 weeks.

To refresh the refrigerated rye sourdough starter before use: Remove from refrigerator 3 days in advance, feeding every day as detailed in the fifth feeding above.

Country Sourdough Bread

DESIRED DOUGH TEMPERATURE: 76°F/24°C

INGREDIENT	US	METRIC	BAKERS' PERCENTAGE
White Sourdough (page 142)	3 lb 5 oz	1.5 kg	30.0%
Rye Sourdough (page 147)	1 lb 2 oz	510 g	10.0%
Water	7 lb 8 oz	3.4 kg	68.0%
Bread flour	6 lb 10 oz	3 kg	60.0%
Whole wheat flour	2 lb 3¼ oz	1 kg	20.0%
Medium rye flour	2 lb 3¼ oz	1 kg	20.0%
Malt syrup	1 oz	30 g	0.5%
Salt	5 oz	140 g	2.8%
Total Dough Weight	23 lb 5 oz	10.6 kg	211.3%

1. Feed the white sourdough starter and rye sourdough starter 18 hours before preparing the final dough.
2. Prepare the final dough a day ahead. Combine the white sourdough, rye sourdough, water, bread flour, whole wheat flour, rye flour, and malt syrup in a mixer and mix on low speed until homogenous, about 3 minutes. Rest the dough in the mixer for 15 to 20 minutes.
3. Add the salt and mix on low speed until the dough reaches the improved stage of gluten development, about 5 minutes.
4. Remove the dough from the mixer and bulk-ferment, covered, for 2 hours. Fold the dough halfway through fermentation. Fold again and ferment a final 15 minutes.
5. Divide the dough into 1 lb 8 oz/680 g pieces and preshape into rounds. Bench-rest the dough for 10 to 15 minutes.
6. Shape the dough into boules and place in lightly floured bannetons, seam side up. Proof the dough for about 1 hour, or as needed, depending on the environment. Place in a refrigerator to retard overnight.
7. Remove the dough from the cooler 1 hour before baking.
8. Score the dough and load into a 475°F/246°C oven. Immediately lower the temperature to 465°F/241°C. Bake until golden brown, about 18 minutes, followed by an additional 15 minutes with the vent open.
9. Place the finished breads on a rack to cool.

Seeded Wheat Sourdough Bread

DESIRED DOUGH TEMPERATURE: 76°F/24°C

INGREDIENT	US	METRIC	BAKERS' PERCENTAGE
Water	8 lb 9½ oz	3.9 kg	65.9%
Bread flour	5 lb 4½ oz	2.4 kg	40.1%
Whole wheat flour	5 lb 12 oz	2.6 kg	44.2%
White Sourdough (page 142)	3 lb 5 oz	1.5 kg	24.5%
Honey	13½ oz	385 g	6.5%
Malt syrup	1 oz	30 g	0.5%
Coarse corn meal	7½ oz	215 g	3.6%
Salt	6¾ oz	170 g	2.8%
Flaxseed	7 oz	200 g	3.4%
Sunflower seeds, toasted	6 oz	170 g	2.7%
Sesame seeds, toasted	6¾ oz	190 g	3.2%
Pumpkin seeds, toasted	6 oz	170 g	2.8%
Total Dough Weight	26 lb 4 oz	11.9 kg	200.7%

1. Make the dough a day ahead. Combine the water, flour, white sourdough, honey, malt syrup, and coarse corn meal in a mixer and mix on low speed until homogenous, about 3 minutes. Stop the mixer and rest the dough to autolyse for about 15 minutes.
2. Resume mixing on low speed for about 1 minute. Add the salt and mix to incorporate, about 30 seconds. Add the seeds and mix until combined, about 1½ minutes.
3. Increase the speed to high and mix until the dough reaches the improved stage of gluten development, about 2 minutes.
4. Remove the dough from the mixer and bulk-ferment, covered, for about 2½ hours. Fold twice during fermentation.
5. Divide the dough into 1 lb 8 oz/680 g pieces and preshape into rounds. Bench-rest the dough for 10 to 15 minutes.
6. Shape the dough into 10-in/25-cm bâtards and place on lightly floured couches. Proof the dough for about 1 hour or as needed, depending on the environment. Transfer to a refrigerator to retard overnight.
7. Bring the dough to room temperature for about 1 hour.
8. Score and load the dough into a 450°F/232°C oven and bake until golden brown, about 30 minutes.
9. Place the finished breads on a rack to cool.

Everything Sourdough Bread

DESIRED DOUGH TEMPERATURE: 76°F/24°C

INGREDIENT	US	METRIC	BAKERS' PERCENTAGE
Sourdough (page 145)	8 lb 13 oz	4 kg	100.0%
Onions, medium diced, sautéed	1 lb 8 oz	680 g	15.0%
Garlic, thinly sliced, sautéed	2¾ oz	75 g	1.8%
Sesame seeds, toasted	3¾ oz	105 g	2.6%
Poppy seeds	3 oz	85 g	2.1%
Cracked black peppercorns	¼ oz	5 g	0.2%
Whole wheat flour	3 oz	85 g	2.1%
Total Dough Weight	11 lb	5 kg	213.6%
Sesame seeds, toasted	As needed	As needed	
Poppy seeds	As needed	As needed	

1. Prepare the dough a day ahead. Mix the sourdough according to the method on page 145. At the end of the mixing, add the onions, garlic, sesame seeds, poppy seeds, pepper, and flour. Mix on low speed for 1 to 2 minutes, until all ingredients are incorporated.
2. Remove the dough from mixer and bulk-ferment, covered, for 2 hours. Fold the dough halfway through fermentation. Fold again and ferment a final 15 minutes.
3. Divide the dough into 1 lb 8 oz/680 gram pieces and preshape into a round. Bench-rest the dough for 10 to 15 minutes.
4. Mix the sesame seeds and poppy seeds. Shape the dough into boules. Roll the boules on a wet towel-lined tray, dip into the seed mixture, and place into lined round bannetons, seam side up.
5. Proof the dough for 30 minutes at room temperature and retard overnight in refrigerator.
6. Remove the dough from the refrigerator 1 hour before baking and allow to come to room temperature.
7. Preheat the hearth oven to 475°F/246°C. Score and load the bread and immediately lower temperature to 465°F/241°C. Bake for 18 minutes, followed by 10 minutes with the vent open.
8. Place the finished breads on a rack to cool.

Apple, Walnut, and Cranberry Sourdough Bread

DESIRED DOUGH TEMPERATURE: 76°F/24°C

INGREDIENT	US	METRIC	BAKERS' PERCENTAGE
Sourdough (page 145)	11 lb 14 oz	5.4 kg	100.0%
Walnuts, toasted	2 lb 14 oz	1.3 kg	21.5%
Apples, diced	2 lb 14 oz	1.3 kg	21.5%
Dried cranberries	1 lb 2 oz	505 g	14.3%
Ground cinnamon	1½ oz	40 g	0.8%
Total Dough Weight	18 lb 13½ oz	8.5 kg	212.5%

1. Make the dough a day ahead. Mix the sourdough according to the method on page 145. At the end of mixing, add the walnuts, apples, and cranberries. Mix on low speed until the ingredients are incorporated, 1 to 2 minutes.
2. Remove the dough from mixer and bulk-ferment, covered, for 2 hours. Fold the dough halfway through fermentation. Fold again and ferment a final 15 minutes.
3. Divide the dough into 1 lb 4 oz/570 g pieces and preshape into rounds. Bench-rest the dough for about 15 minutes.
4. Shape the dough into 10-in/25-cm bâtards and place into lightly floured bannetons. Proof for about 1 hour, or as needed, depending on the environment. Place the dough in a refrigerator to retard overnight.
5. Remove the dough from the cooler 1 hour before baking.
6. Score the dough and load into a 475°F/246°C oven. Immediately lower the temperature to 465°F/241°C. Bake until golden brown, about 18 minutes, followed by an additional 15 minutes with the vent open.
7. Place the finished breads on a rack to cool.

VARIATION

For Herb Sourdough Bread, omit the walnuts, apples, cranberries, and cinnamon. Add ½ oz/14 g each chopped thyme, chopped rosemary, basil chiffonade, and 1 oz/30 g minced garlic. Proceed with Step 2 in the recipe.

Chipotle Sourdough Bread

DESIRED DOUGH TEMPERATURE: 76°F/24°C

INGREDIENT	US	METRIC	BAKERS' PERCENTAGE
Sourdough (page 145)	10 lb 2½ oz	4.6 kg	100.0%
Chipotles in adobo, puréed	9 oz	260 g	5.7%
Whole wheat flour	5 oz	140 g	2.9%
Total Dough Weight	11 lb 2½ oz	5 kg	108.6%

1. A day ahead, prepare the sourdough according to the method on page 145. At the end of the mixing, add the chipotles and flour. Mix on low speed until the ingredients are incorporated, 2 to 3 minutes.
2. Remove the dough from the mixer and bulk-ferment for 2 hours. Fold the dough halfway through fermentation. Fold again and ferment a final 15 minutes.
3. Divide the dough into 1 lb 8 oz/680 g pieces and preshape into rounds. Bench-rest the dough for 10 to 15 minutes.
4. Shape the dough into boules and place in lightly floured bannetons, seam side up. Proof for about 1 hour or as needed, depending on the environment. Place in a refrigerator to retard overnight.
5. Remove the dough from the refrigerator 1 hour before baking.
6. Score the dough and load into a 475°F/246°C oven. Immediately lower the temperature to 465°F/241°C. Bake until golden brown, about 18 minutes, followed by an additional 15 minutes with the vent open.
7. Place the finished breads on a rack to cool.

Roasted Garlic and Jack Cheese Sourdough Bread

DESIRED DOUGH TEMPERATURE: 76°F/24°C

INGREDIENT	US	METRIC	BAKERS' PERCENTAGE
Sourdough (page 145)	8 lb 2½ oz	3.7 kg	100.0%
Garlic, roasted	5½ oz	155 g	4.2%
Monterey Jack cheese, cubed	2 lb 3¼ oz	1 kg	29.2%
Whole wheat flour	2¾ oz	75 g	2.1%
Total Dough Weight	11 lb	5 kg	135.4%

1. A day ahead, mix the sourdough according to the method on page 145. At the end of the mixing, add the garlic, cheese, and flour. Mix on low speed for 1 to 2 minutes, until all ingredients are incorporated.
2. Remove the dough from the mixer and allow to bulk-ferment, covered, for 1 hour. Fold and ferment another 1 hour. Fold a final time and ferment an additional 15 minutes.
3. Divide the dough into 1 lb 4 oz/570 g pieces and preshape into a round. Bench-rest the dough for 10 to 15 minutes.
4. Shape the dough into 10-in/25-cm bâtards and place in lightly floured banneton molds. Cover and ferment for 1 hour at room temperature. Retard overnight in the refrigerator.
5. Remove the dough from the refrigerator 1 hour before baking and allow to come to room temperature.
6. Preheat the hearth oven to 475°F/246°C. Score and load the dough and immediately lower temperature to 465°F/241°C. Bake for 18 minutes, followed by an additional 15 minutes with the vent open, until golden brown.
7. Place the finished breads on a rack to cool.

Sourdough Bread with Mushrooms

DESIRED DOUGH TEMPERATURE: 76°F/24°C

INGREDIENT	US	METRIC	BAKERS' PERCENTAGE
White mushrooms, sliced	1 lb 5 oz	595 g	15.0%
Garlic, minced	2 oz	60 g	1.5%
Onions, medium diced	7½ oz	215 g	5.4%
Olive oil	1 oz	30 g	0.7%
Thyme, minced	2 oz	60 g	1.5%
Sourdough (page 145)	8 lb 9½ oz	3.9 kg	100.0%
Total Dough Weight	10 lb 13¾ oz	4.9 kg	126.2%

1. A day ahead, sauté the mushrooms, garlic, and onions in the olive oil. Add the thyme and mix well. Drain off any excess liquid.
2. Mix the sourdough according to the method on page 145. Add the mushroom mixture and mix on low speed until incorporated, 1 to 2 minutes.
3. Remove the dough from mixer and bulk-ferment for 2 hours. Fold once halfway through fermentation. Fold again and ferment an additional 15 minutes.
4. Divide the dough into 1 lb 8 oz/680 g pieces and preshape into rounds. Bench-rest the dough for 10 to 15 minutes.
5. Shape the dough into boules and place seam side up into floured and lined round bannetons. Cover with plastic. Proof for 1 hour at room temperature or as needed, depending on the environment. Place in the refrigerator to retard overnight.
6. Remove the dough from the refrigerator 1 hour before baking.
7. Score and load into a 475°F/246°C oven and immediately lower the temperature to 465°F/241°C. Bake until golden brown, about 18 minutes, followed by an additional 15 minutes with the vent open.
8. Place the finished breads on a rack to cool.

VARIATIONS

For a Jalapeño and Cheddar Cheese Sourdough Bread, replace the sautéed mushrooms, garlic, onion, olive oil, and thyme with 8¼ oz/235 g pickled jalapeños (drained) and 2 lb 5¼ oz/1.1 kg diced Cheddar cheese. Add the ingredients at Step 2 and continue as described.

For an Orange, Currant, and Almond Sourdough Bread, replace the sautéed mushrooms, garlic, onion, olive oil, and thyme with 14 oz/400 g toasted slivered almonds, 8¼ oz/240 g dried currants, and 1 lb 1 oz/475 g candied orange peel (drained). Add the ingredients at Step 2 and continue as described.

Sun-dried Tomato and Asiago Sourdough Bread

DESIRED DOUGH TEMPERATURE: 76°F/24°C

INGREDIENT	US	METRIC	BAKERS' PERCENTAGE
Sourdough (page 145)	8 lb 9½ oz	3.9 kg	100.0%
Sun-dried tomatoes	1 lb 7 oz	650 g	16.7%
Asiago, grated	11½ oz	325 g	8.3%
Whole wheat flour	4½ oz	130 g	3.3%
Total Dough Weight	11 lb	5 kg	128.3%
Asiago, grated	As needed	As needed	

1. A day ahead, mix the sourdough according to the method on page 145. Add the tomatoes, cheese, and flour. Mix on low speed until incorporated, 1 to 2 minutes.
2. Remove the dough from the mixer and bulk-ferment for 2 hours. Fold halfway through the fermentation. Fold again and ferment an additional 15 minutes.
3. Divide the dough into 1 lb 8 oz/680 g pieces and preshape into a round. Bench-rest the dough for 10 to 15 minutes.
4. Shape the dough into 14-in/36-cm bâtards and place in a lightly floured couche. Cover with plastic. Proof the dough for 1 hour or as needed, depending on the environment. Place in the refrigerator to retard overnight.
5. Remove the dough from the refrigerator 1 hour before baking and bring to room temperature.
6. Place the dough on a silicone mat or parchment paper, score the loaves, and load into a 475°F/246°C oven. Immediately lower the temperature to 450°F/232°C, and bake until golden brown, about 18 minutes.
7. Briefly remove the dough from the oven, spray with water, and top with the remaining cheese (1 oz/30 g per loaf). Return to the oven and bake with the vent open, until the crust is developed and the cheese is melted and bubbly, about 15 minutes.
8. Place the finished breads on a rack to cool.

Durum Sourdough Bread

DESIRED DOUGH TEMPERATURE: 78°F/26°C

INGREDIENT	US	METRIC	BAKERS' PERCENTAGE
Water	9 lb 14½ oz	4.5 kg	90.0%
Durum flour	5 lb 8 oz	2.5 kg	50.0%
Bread flour	5 lb 8 oz	2.5 kg	50.0%
Durum Sour (page 204)	4 lb	1.8 kg	35.0%
Yeast, instant dry	¾ oz	20 g	0.4%
Salt	4½ oz	130 g	2.6%
Total Dough Weight	25 lb 3¾ oz	11.45 kg	228.0%

1. Combine the water, durum flour, bread flour, and durum sour in a mixer and mix on low speed until homogenous, about 2 minutes. Rest the dough in the mixer to autolyse for 15 to 20 minutes.
2. Add the yeast and mix on low speed to incorporate, about 2½ minutes. Add the salt and mix to combine, about 30 seconds. Increase the speed to high and mix until the dough reaches the improved stage of gluten development, about 2 minutes.
3. Bulk-ferment, covered, until doubled in size, about 2 hours. Fold the dough every 30 minutes.
4. Divide the dough into 1 lb 8 oz/680 g pieces and preshape into rounds. Bench-rest the dough for 15 to 20 minutes.
5. Shape the dough into boules and place in durum-dusted round bannetons. Proof the dough for 60 to 90 minutes or as needed, depending on the environment.
6. Score the dough and load into a 470°F/243°C oven. After 10 minutes, reduce the temperature to 450°F/232°C and bake until golden brown, about 20 minutes. Open the vent and bake an additional 10 minutes.
7. Place finished breads on a rack to cool.

Family-of-Four Sourdough Bread

DESIRED DOUGH TEMPERATURE: 76°F/24°C

INGREDIENT	US	METRIC	BAKERS' PERCENTAGE
White Sourdough (page 142)	1 lb 14 oz	860 g	30.0%
Rye Sourdough (page 145)	8½ oz	240 g	10.0%
Water	3 lb 8½ oz	1.6 kg	68.0%
Bread flour	3 lb 5 oz	1.5 kg	60.0%
Whole wheat flour	12½ oz	355 g	15.0%
Spelt flour	12½ oz	355 g	15.0%
Medium rye flour	8½ oz	240 g	10.0%
Malt syrup	½ oz	15 g	0.5%
Salt	2½ oz	70 g	2.8%
Total Dough Weight	11 lb	5 kg	211.3%

1. A day ahead, combine the white sourdough, rye sourdough, water, bread flour, whole wheat flour, spelt flour, rye flour, and malt syrup in a mixer. Mix on low speed until combined, about 3 minutes. Allow the dough to rest in the mixer for 15 to 20 minutes.
2. Add the salt and mix on low speed until it reaches the improved stage of gluten development, about 5 minutes.
3. Remove the dough from the mixer and bulk-ferment, covered, until it has doubled in size, about 2 hours. Fold once halfway through fermentation. Fold a second time and ferment for an additional 15 minutes.
4. Divide the dough into 1 lb 8 oz/680 g pieces and preshape into rounds. Bench-rest the dough for about 15 minutes.
5. Shape the dough into boules and place seam side up in lightly floured bannetons. Proof at room temperature for about 1 hour. Retard overnight in a refrigerator.
6. Remove from the refrigerator 1 hour before baking.
7. Score the dough and load into a 475°F/246°C oven. Immediately lower the temperature to 465°F/241°C and bake for 18 minutes. Open the vent and bake for an additional 15 minutes, until golden brown.
8. Place the finished breads on a rack to cool.

Semolina Sourdough Bread

DESIRED DOUGH TEMPERATURE: 78°F/26°C

INGREDIENT	US	METRIC	BAKERS' PERCENTAGE
Water	7 lb	3.2 kg	64.3%
Bread flour	2 lb 2½ oz	975 g	19.5%
Semolina flour	4 lb 6½ oz	2 kg	40.2%
Durum flour	4 lb 6½ oz	2 kg	40.2%
Liquid Levain (page 132)	5 lb 8 oz	2.5 kg	50.3%
Yeast, instant dry	½ oz	15 g	0.3%
Salt	5½ oz	160 g	3.2%
Total Dough Weight	23 lb 13½ oz	10.9 kg	218.0%

1. A day ahead, combine the water, bread flour, semolina flour, liquid white sour, and yeast in a mixer. Mix on low speed until homogenous. Rest the dough in the mixer to autolyse for 10 to 15 minutes.
2. Add the salt and mix on low speed until the dough reaches the improved stage of gluten development, about 3 minutes.
3. Remove the dough from the mixer and bulk-ferment for 2 hours. Fold the dough halfway through fermentation. Fold the dough again and ferment for an additional 10 minutes.
4. Divide the dough into 1 lb 8 oz/680 g pieces and preshape into rounds. Bench-rest the dough for 10 to 15 minutes.
5. Shape the dough into 10-in/25-cm bâtards and place on a lightly floured couche. Keep the dough covered as you work. Proof the dough for 1 hour at room temperature or as needed. Place in the refrigerator to retard overnight.
6. Remove the dough from the refrigerator 1 hour before baking.
7. Score the bread and load into a 475°F/246°C oven. Immediately lower the temperature to 465°F/241°C. Bake for 3 minutes, and then steam the oven. Bake for an additional 15 minutes, until golden brown. Open the vent, and bake for 10 more minutes.
8. Place the finished breads on a rack to cool.

Pane alle Olive

DESIRED DOUGH TEMPERATURE: 80°F/27°C

INGREDIENT	US	METRIC	BAKERS' PERCENTAGE
Biga, 85% hydration (page 125)	1 lb 10½ oz	750 g	15.0%
Water	6 lb 2½ oz	2.8 kg	55.0%
Bread flour	11 lb	5 kg	100.0%
Malt syrup	1 lb 10½ oz	750 g	15.0%
Yeast, instant dry	3 oz	90 g	1.8%
Sea salt	3½ oz	100 g	2.0%
Olive oil, good quality	1 lb 10½ oz	750 g	15.0%
Black Ligurian olives, pitted	3½ oz	100 g	2.0%
Total Dough Weight	22 lb 12 oz	10.3 kg	205.8%

1. Prepare the biga the day before mixing the final dough.
2. Combine the biga, water, flour, and malt syrup in a mixer, and mix on low speed until homogenous, about 3 minutes. Sprinkle the yeast on top of the dough and rest in the mixer to autolyse, 10 to 15 minutes.
3. Resume mixing on low speed to incorporate the yeast, about 1 minute. Add the salt and mix to combine, about 30 seconds. Increase the speed to high and mix until the dough reaches the improved stage of gluten development, about 3 minutes.
4. Return the mixer to low speed, stream in the olive oil, and add the olives. Mix until combined, about 1 minute.
5. Remove the dough from the mixer and bulk-ferment, covered, until doubled in size, about 2 hours. Fold once halfway through fermentation.
6. Divide the dough into 1 lb 10½ oz/750 g pieces and preshape into rounds. Bench-rest the dough for about 15 minutes.
7. Shape into rounds and place seam side up in floured round bannetons. Proof the dough for 1 to 1½ hours or as needed, depending on the environment.
8. Score and load the dough into a 450°F/232°C oven, steaming for 1 second before loading and 3 seconds after loading. Bake until dark golden brown, about 50 minutes.
9. Place the finished breads on a rack to cool.

Pain au Levain

DESIRED DOUGH TEMPERATURE: 76°F/24°C

INGREDIENT	US	METRIC	BAKERS' PERCENTAGE
Water	7 lb 8 oz	3.4 kg	67.0%
Bread flour	8 lb 6 oz	3.8 kg	75.0%
Whole wheat flour	2 lb 14 oz	1.3 kg	25.0%
Liquid Levain (page 132)	4 lb	1.8 kg	36.9%
Sea salt	3¾ oz	105 g	2.1%
Total Dough Weight	23 lb	10.4 kg	206.0%

1. Combine the water, bread flour, whole wheat flour, and liquid levain in a mixer. Mix on low speed until homogenous, about 1 minute. Stop the mixer and rest the dough to autolyse, about 15 minutes.
2. Resume mixing on low speed for an additional minute. Add the salt and mix for 30 seconds, then increase the speed to high. Mix on high speed until the dough reaches the improved stage of gluten development, about 2 minutes.
3. Bulk-ferment, covered, for about 2 hours, folding the dough once halfway through fermentation.
4. Divide the dough into 1 lb 10 oz/750 g pieces and preshape into boules. Bench-rest the dough for about 15 minutes.
5. Shape into boules and place seam side up in floured bannetons. Proof the dough for about 2 hours or as needed, depending on the environment.
6. Dust the dough with flour and score. Load the dough into a 470°F/243°C deck oven. Steam 1 second before loading and 3 seconds after loading. Bake until dark golden brown, about 45 minutes.
7. Place the finished breads on a rack to cool.

Pane al'Olio

DESIRED DOUGH TEMPERATURE: 80°F/27°C

INGREDIENT	US	METRIC	BAKERS' PERCENTAGE
Biga, 85% hydration (page 125)	2 lb 14 oz	1.3 kg	50.0%
Water	3 lb 1½ oz	1.4 kg	55.0%
Bread flour	2 lb 8 oz	2.5 kg	100.0%
Yeast, instant dry	2¾ oz	75 g	3.0%
Sea salt	2 oz	55 g	2.2%
Olive oil	4½ oz	125 g	5.0%
Total Dough Weight	9 lb	5.5 kg	215.2%

1. Mix the biga the day before preparing the final dough.
2. Combine the biga, water, and flour in a mixer. Mix on low speed until homogenous, 1 to 2 minutes.
3. Sprinkle the yeast on top of the dough and rest in the mixer to autolyse, 10 to 15 minutes.
4. Mix on low speed to incorporate the yeast, about 30 seconds. Add the salt and mix for an additional 30 seconds, or until incorporated.
5. Increase the speed to high and slowly add the olive oil. Mix until the dough reaches the improved stage of gluten development, about 4 minutes.
6. Remove the dough from the mixer and bulk-ferment, covered, until doubled in size, about 2 hours. Fold once halfway through fermentation.
7. Divide the dough into 3 oz/85 g pieces and pre-shape into rounds. Bench-rest the dough for about 15 minutes, until relaxed.
8. Shape the dough into rounds and place seam side down onto floured couches or in rings on parchment-lined sheet trays. Proof the dough for about 1½ hours or as needed, depending on the environment.
9. Score the dough and load into a 450°F/232°C oven, steaming 1 second before loading and 3 seconds after loading. Bake until golden brown, about 18 minutes.
10. Place the finished breads on a rack to cool.

Wheat Bread

DESIRED DOUGH TEMPERATURE: 78°F/26°C

INGREDIENT	US	METRIC	BAKERS' PERCENTAGE
Bread flour	6 lb 10 oz	3 kg	60.0%
Whole wheat flour	4 lb 6½ oz	2 kg	40.0%
Yeast, instant dry	1½ oz	35 g	0.7%
Salt	4 oz	110 g	2.2%
Sugar	7 oz	200 g	4.0%
Oil	12 oz	340 g	6.7%
Malt syrup	¾ oz	20 g	0.4%
Milk	7 lb 11 oz	3.5 kg	69.3%
Total Dough Weight	20 lb 4¾ oz	9.2 kg	183.3%

1. Combine the bread flour, whole wheat flour, yeast, salt, sugar, oil, malt syrup, and milk in a mixer, and mix on low speed until combined, about 4 minutes. Increase the speed to high and mix until the dough reaches the improved stage of gluten development, about 4 minutes.
2. Remove the dough from the mixer and bulk-ferment, covered, until doubled in size, about 45 minutes.
3. Divide the dough into 2 lb 7 oz/1.1 kg pieces and preshape into 10-in/25-cm oblong loaves. Bench-rest the dough for 10 to 15 minutes.
4. Shape the dough into 18-in/46-cm oblong loaves and place into oiled Pullman pans. Proof for 1 hour with the pan lids off, or as needed, depending on the environment.
5. Place the oiled lids onto the pans and load into a 400°F/204°C convection oven. Immediately lower the temperature to 375°F/191°C and bake until golden brown, about 45 minutes.
6. Immediately remove the baked loaves from the pans. Place the finished breads on a rack to cool.

Whole Wheat Bread with Biga

DESIRED DOUGH TEMPERATURE: 78°F/26°C

INGREDIENT	US	METRIC	BAKERS' PERCENTAGE
Biga			
Water	15¼ oz	435 g	8.7%
Whole wheat flour	14 oz	395 g	7.9%
Bread flour	14 oz	395 g	7.9%
Yeast, instant dry	¼ tsp	1 g	–
Final Dough			
Water	7 lb 1 oz	3.2 kg	63.8%
Whole wheat flour	6 lb 6 oz	2.9 kg	58.0%
Bread flour	2 lb 14 oz	1.3 kg	26.1%
Yeast, instant dry	¾ oz	20 g	0.4%
Honey	6½ oz	185 g	3.7%
Malt syrup	¾ oz	20 g	0.4%
Biga	2 lb 10 oz	1.2 kg	24.6%
Salt	4 oz	115 g	2.3%
Total Dough Weight	19 lb 11 oz	8.9 kg	179.3%

1. Prepare the biga 18 hours before mixing the final dough (see page 125).
2. To prepare the final dough, combine the water, whole wheat flour, bread flour, yeast, honey, malt syrup, and biga in a mixer. Mix on low speed until homogenous, about 3 minutes. Allow the dough to rest in the mixer for 15 minutes to autolyse.
3. Add the salt and mix on low speed until the dough reaches the improved stage of gluten development, about 3 minutes.
4. Remove the dough from the mixer and bulk-ferment, covered, for 45 minutes. Fold the dough and ferment an additional 15 minutes.
5. Divide the dough into 1 lb 4 oz/570 g pieces and preshape into rounds. Bench-rest the dough for 10 to 15 minutes.
6. Shape the dough into 12-in/30-cm oblong loaves and place seam side up onto lightly floured couches. Proof for 30 to 40 minutes or as needed, depending on the environment.
7. Score the dough and load into a 470°F/243°C oven. Immediately lower the temperature to 440°F/227°C. Bake until golden brown, about 18 minutes, followed by an additional 8 minutes with the vent open.
8. Place the finished breads on a rack to cool.

Whole Wheat Bread with Poolish

DESIRED DOUGH TEMPERATURE: 78°F/26°C

INGREDIENT	US	METRIC	BAKERS' PERCENTAGE
Poolish			
Water	3 lb 12 oz	1.7 kg	33.3%
Whole wheat flour	3 lb 12 oz	1.7 kg	33.3%
Yeast, instant dry	¼ tsp	1 g	—
Final Dough			
Water	4 lb 3 oz	1.9 kg	37.6%
Bread flour	3 lb 12 oz	1.7 kg	33.3%
Whole wheat flour	3 lb 12 oz	1.7 kg	33.3%
Yeast, instant dry	¾ oz	20 g	0.4%
Salt	4 oz	115 g	2.3%
Poolish	7 lb 4½ oz	3.3 kg	66.7%
Total Dough Weight	19 lb 4¼ oz	8.7 kg	173.6%

1. Prepare the poolish 18 hours before mixing the final dough (see page 124).
2. To prepare the final dough, combine the water, bread flour, whole wheat flour, yeast, salt, and poolish in a mixer, and mix on low speed until homogenous, about 4 minutes. Increase the speed to high and mix until the dough reaches the intense stage of gluten development, about 2 minutes.
3. Remove the dough from the mixer and ferment for 1 hour. Fold the dough and ferment for an additional 15 minutes.
4. Divide the dough into 1 lb 4 oz/570 g pieces and preshape into rounds. Bench-rest the dough for 10 to 15 minutes.
5. Shape the rounds into 10-in/25-cm oblong loaves and place seam side up onto lightly floured couches. Proof the dough for 30 to 40 minutes or as needed, depending on the environment.
6. Score the dough and load into a 450°F/232°C oven and immediately lower the temperature to 440°F/227°C. Bake until golden brown, about 15 minutes, followed by an additional 15 minutes with the vent open.
7. Place the finished breads on a rack to cool.

Whole Wheat Bread with Sesame and Asiago

DESIRED DOUGH TEMPERATURE: 78°F/26°C

INGREDIENT	US	METRIC	BAKERS' PERCENTAGE
Biga			
Water	1 lb 9½ oz	725 g	14.5%
Whole wheat flour	1 lb 15 oz	875 g	17.5%
Bread flour	13¼ oz	375 g	7.5%
Yeast, instant dry	¼ tsp	1 g	—
Final Dough			
Water	6 lb 6 oz	2.9 kg	58.3%
Whole wheat flour	5 lb 12 oz	2.6 kg	52.5%
Bread flour	2 lb 7 oz	1.1 kg	22.5%
Yeast, instant dry	¾ oz	20 g	0.4%
Salt	4 oz	115 g	2.3%
Malt syrup	1 oz	30 g	0.5%
Biga	4 lb 3 oz	1.9 kg	39.5%
Asiago, diced	2 lb 10 oz	1.2 kg	23.4%
Sesame seeds, toasted	8 oz	230 g	4.5 %
Total Dough Weight	22 lb 10 oz	10 kg	203.9%
Sesame seeds	As needed	As needed	
Asiago, grated	As needed	As needed	

1. Mix the biga 18 hours before preparing the final dough (page 125).
2. To prepare the final dough, combine the water, whole wheat flour, bread flour, yeast, salt, malt syrup, and biga in a mixer. Mix on low speed until homogenous, about 4 minutes. Increase the speed to high and mix until the dough reaches the intense stage of gluten development, about 2 minutes.
3. Add the cheese and sesame seeds, and mix on low speed until incorporated, about 1 minute. Increase the speed and mix on high speed for 30 seconds.
4. Remove the dough from the mixer and bulk-ferment for 45 minutes. Fold the dough and ferment for an additional 15 minutes.
5. Divide the dough into 1 lb 4 oz/570 g pieces and preshape into rounds. Bench-rest the dough for 10 to 15 minutes.
6. Shape the dough into 12-in/30-cm oblong loaves and dip the tops into sesame seeds. Place onto a couche. Proof the dough for 30 to 40 minutes or as needed, depending on the environment.
7. Place the dough on a silicone mat or parchment paper and score. Load into a 465°F/241°C oven and immediately lower the temperature to 450°F/232°C. Bake until golden brown, about 20 minutes.
8. Briefly remove the breads from the oven and top with the grated cheese (about 1 oz/30 g per loaf). Bake with the vent open until the crust is developed and the cheese is bubbly, about 8 minutes.
9. Place the finished breads on a rack to cool.

Soft Multigrain Bread

DESIRED DOUGH TEMPERATURE: 78°F/26°C

INGREDIENT	US	METRIC	BAKERS' PERCENTAGE
Milk Soaker			
Milk	1 lb 10½ oz	750 g	15.6%
Flaxseed	4½ oz	125 g	2.6%
Sesame seeds, toasted	¼ oz	10 g	0.1%
9-grain mix	1 lb 6 oz	625 g	13.0%
Final Dough			
Bread flour	7 lb 8 oz	3.4 kg	71.3%
Whole wheat flour	1 lb 6 oz	625 g	13.0%
Yeast, instant dry	1½ oz	45 g	0.9%
Salt	4½ oz	130 g	2.7%
Sunflower oil	¾ oz	20 g	0.4%
Malt syrup	1 oz	30 g	0.5%
Milk	5 lb 4½ oz	2.4 kg	49.9%
Honey	1 lb 9 oz	705 g	14.7%
Milk Soaker	3 lb 12 oz	1.7 kg	35.2%
Total Dough Weight	20 lb	9.1 kg	188.6%

1. Prepare the milk soaker 18 hours before mixing the final dough. Combine the milk, flaxseed, sesame seeds, and 9-grain mix in a container. Cover and store at room temperature.
2. To prepare the final dough, combine the bread flour, whole wheat flour, yeast, salt, oil, malt syrup, milk, and honey in a mixer. Mix on low speed until homogenous, about 4 minutes. Increase the speed to high and mix until the dough reaches the intense stage of gluten development, about 4 minutes.
3. Add half the soaker and mix on low speed until incorporated, about 1 minute. Add the remaining soaker and mix until combined, about 1 minute.
4. Remove the dough from the mixer and bulk-ferment, covered, until doubled in size, about 45 minutes.
5. Divide the dough into 3 lb/1.4 kg pieces and pre-shape into oblong loaves. Bench-rest the dough for 10 to 15 minutes.

6. Shape the dough into 18-in/46-cm oblong loaves and place into oiled Pullman pans (see Note). Proof, without the lids, for 45 minutes or as needed, depending on the environment.

7. Place the oiled lids on the pans and load into a 470°F/243°C convection oven. Immediately lower the temperature to 410°F/210°C and bake until golden brown, about 16 minutes.

8. Immediately remove the baked loaves from the pans. Place the finished breads on a rack to cool.

CHEF'S NOTE

This dough can also be used to make rolls. Divide the dough into 4 lb/1.8 kg presses and divide on a dough divider. Place on a parchment-lined sheet tray, and brush with egg wash before and after proofing. Proceed with Step 5 in the recipe.

Multigrain Bread

DESIRED DOUGH TEMPERATURE: 76°F/24°C

INGREDIENT	US	METRIC	BAKERS' PERCENTAGE
Grain Soaker			
Water, warmed	3 lb 12 oz	1.7 kg	26.6%
9-grain mix	2 lb 3¼ oz	1 kg	15.6%
Flaxseed	5 oz	140 g	2.2%
Sunflower seeds	9 oz	260 g	3.9%
Final Dough			
Water	6 lb 2½ oz	2.8 kg	44.1%
Bread flour	6 lb 6 oz	2.9 kg	45.1%
Whole wheat flour	4 lb 10 oz	2.1 kg	33.2%
Salt	5½ oz	150 g	2.3%
Yeast, instant dry	1½ oz	40 g	0.6%
Pâte Fermentée (page 199)	5 lb 8 oz	2.5 kg	39.1%
Malt syrup	1½ oz	40 g	0.6%
Grain Soaker	6 lb 13 oz	3.1 kg	48.4%
Total Dough Weight	30 lb	13.6 kg	213.4%
Sesame seeds	As needed	As needed	

1. Prepare the grain soaker 18 hours before mixing the final dough. Combine the warm water, 9-grain mix, flaxseed, and sunflower seeds in a covered container. Store at room temperature.
2. To prepare the final dough, combine the water, bread flour, whole wheat flour, salt, yeast, pâte fermentée, and malt syrup in a mixer. Mix on low speed until homogenous, about 3 minutes. Increase the speed to high and mix for about 2 minutes. Add half the soaker and mix on low speed until combined, about 2 minutes. Add the remaining soaker and mix until incorporated, about 2 minutes.
3. Increase the speed to high and mix until the dough reaches the intense stage of gluten development, about 1 minute.
4. Remove the dough from the mixer and bulk-ferment for about 30 minutes. Fold the dough and ferment for an additional 15 minutes.

5. Divide the dough into 1 lb 4 oz/570 g pieces and preshape into rounds. Bench-rest the dough for 10 to 15 minutes.
6. Shape the dough into 10-in/25-cm bâtards. Roll the loaf on a tray lined with wet towels, sprinkle with the sesame seeds, and place into lined bannetons. Proof the dough for 30 to 35 minutes or as needed, depending on the environment.
7. Score and load the dough into a 475°F/246°C oven and immediately lower the temperature to 465°F/241°C. Bake until lightly golden, about 20 minutes. Open the vent and bake until golden brown, about 12 minutes.
8. Place the finished breads on a rack to cool.

Alternative Grain Bread

DESIRED DOUGH TEMPERATURE: 78°F/26°C

INGREDIENT	US	METRIC	BAKERS' PERCENTAGE
Soaker			
Water, room temperature	4 lb 13-½ oz	2.2 kg	43.2%
Flaxseed	1 lb 6 oz	635 g	11.4%
Quinoa	1 lb 6 oz	635 g	11.4%
Amaranth	1 lb 6 oz	635 g	11.4%
Teff	12 oz	345 g	6.9%
Final Dough			
Water	4 lb 6½ oz	2.1 kg	39.9%
Kamut flour	2 lb 6 oz	1.2 kg	21.6%
Spelt flour	4 lb 7½ oz	2.1 kg	40.5%
Bread flour	4 lb 3 oz	1.9 kg	37.9%
Salt	5½ oz	155 g	3.1%
Yeast, instant dry	1 oz	30 g	0.7%
Malt syrup	½ oz	15 g	0.4 %
Honey	14½ oz	415 g	8.2%
Pâte Fermentée (page 199)	5 lb 15½ oz	2.3 kg	54.1%
White Sourdough (page 142)	2 lb 6 oz	1.2 kg	21.6%
Soaker	9 lb 4½ oz	4.3 kg	84.2%
Total Dough Weight	34 lb 6½ oz	15.7 kg	312.2%

1. Prepare the soaker 18 hours before mixing the final dough (see page 131).
2. To prepare the final dough, combine the water, Kamut flour, spelt flour, bread flour, salt, yeast, malt syrup, honey, pâte fermentée, and white sourdough in a mixer. Mix on low speed until homogenous, about 5 minutes. Increase the speed to high and mix for an additional 3 minutes, until the dough is smooth and well incorporated.
3. Add half the soaker to the dough, and mix on low speed for 1 minute. Add the remaining soaker and mix for an additional 1 minute on low speed, just until the soaker is distributed.
4. Increase the speed, and mix on high until the dough reaches the improved stage of gluten development, about 2 minutes.

5. Remove the dough from the mixer and bulk-ferment, covered, until it has doubled in size, about 1 hour. Fold once halfway through fermentation.
6. Divide the dough into 1 lb 12 oz/795 g pieces and preshape into rounds. Bench-rest the dough for about 15 minutes.
7. Shape the dough into 10-in/25-cm bâtards and place into greased strap pans. Proof for about 1 hour or as needed, depending on the environment.
8. Load the pans into a 450°F/232°C oven. Immediately steam the oven and lower the temperature to 400°F/204°C. Bake until dark golden brown, 30 to 40 minutes.
9. Place the finished breads on a rack to cool.

Flaxseed and Spelt Multigrain Bread

DESIRED DOUGH TEMPERATURE: 78°F/26°C

INGREDIENT	US	METRIC	BAKERS' PERCENTAGE
Grain Soaker			
Water, boiling	5 lb 12 oz	2.6 kg	35.3%
Cracked rye berries	1 lb 8 oz	685 g	8.1%
Cracked wheat berries	1 lb 8 oz	685 g	8.1%
Flaxseed	2 lb 1 oz	950 g	16.2%
Final Dough			
Water	5 lb 8 oz	2.5 kg	34.3%
Bread flour	4 lb 6½ oz	2 kg	27.0%
Medium rye flour	3 lb 1½ oz	1.4 kg	18.7%
Spelt flour	3 lb 8½ oz	1.6 kg	21.9%
Salt	6¾ oz	195 g	2.6%
Yeast, instant dry	2 oz	55 g	0.7%
Rye Sourdough (page 145)	5 lb 1 oz	2.3 kg	30.6%
Malt syrup	1½ oz	35 g	0.5%
Grain Soaker	11 lb	5.0 kg	67.7%
Total Dough Weight	33 lb 3¾ oz	15.1 kg	204.0%
Sesame seeds	As needed	As needed	

1. Prepare the grain soaker a day before mixing the final dough. Pour the water over the cracked rye and wheat berries and the flaxseed. Cover and store at room temperature for 18 hours.
2. To prepare the final dough, combine the water, bread flour, rye flour, spelt flour, salt, yeast, rye sour, and malt syrup in a mixer. Mix on low speed until homogenous, about 3 minutes. Add half the grain soaker and mix until incorporated, about 2 minutes. Add the remaining soaker and mix until combined, about 2 minutes.
3. Increase the speed to high and mix until the dough reaches the intense stage of gluten development, about 3 minutes.
4. Remove the dough from the mixer and bulk-ferment, covered, for about 30 minutes.

5. Divide the dough into 1 lb 4 oz/570 g pieces and preshape into rounds. Bench-rest the dough for 10 to 15 minutes.
6. Shape the dough into 10-in/25-cm bâtards. Roll the loaves on a tray lined with wet towels, coat with sesame seeds, and place into lined bannetons. Proof the dough for 30 to 35 minutes or as needed, depending on the environment.
7. Score and load into a 485°F/252°C oven and immediately lower the temperature to 475°F/246°C. Bake until lightly golden, about 18 minutes. Open the vent and bake until golden brown, about 12 minutes.
8. Place the finished breads on a rack to cool.

Semolina Bread

DESIRED DOUGH TEMPERATURE: 78°F/26°C

INGREDIENT	US	METRIC	BAKERS' PERCENTAGE
Biga			
Water	2 lb ½ oz	920 g	18.3%
Durum flour	1 lb 6 oz	630 g	12.6%
Salt	4 oz	115 g	2.3%
Sugar	4½ oz	130 g	2.6%
Yeast, instant dry	1 oz	30 g	0.6%
Final Dough			
Water	6 lb	2.7 kg	53.7%
Biga	5 lb 12 oz	2.6 kg	51.7%
Olive oil	7½ oz	210 g	4.2%
Malt syrup	1 oz	30 g	0.5%
Durum flour	1 lb 13½ oz	835 g	16.7%
Semolina flour	1 lb 13½ oz	835 g	16.7%
Bread flour	3 lb 12 oz	1.7 kg	33.3%
Yeast, instant dry	¾ oz	25 g	0.5%
Salt	4 oz	110 g	2.2%
Total Dough Weight	20 lb	9 kg	69.4%
Sesame seeds or semolina flour	As needed	As needed	

1. Prepare the biga 18 hours before mixing the final dough (see page 125).
2. To prepare the final dough, combine the water, biga, oil, malt syrup, durum flour, semolina flour, bread flour, and yeast in a mixer, and mix on low speed until homogenous, about 3 minutes. Cover the dough in the mixer and rest for 10 to 15 minutes to autolyse.
3. Mix the dough on low speed for about 2 minutes, and add the salt. Increase the speed to high and mix until the dough reaches the improved stage of gluten development, about 2 minutes.
4. Remove the dough from the mixer and bulk-ferment, covered, for about 45 minutes. Fold and ferment for an additional 15 minutes.

5. Divide the dough into 1 lb 2 oz/510 g pieces and preshape into logs. Bench-rest the dough for 10 to 15 minutes.

6. Shape the dough into 18-in/46-cm baguettes and roll the tops in semolina flour or sesame seeds. Place on a couche. Proof the dough for 30 to 40 minutes or as needed, depending on the environment.

7. Score and load the dough into a 470°F/243°C oven and bake for 15 minutes until lightly browned. Open the vent and bake until golden brown, about 8 minutes.

8. Place the finished breads on a rack to cool.

Semolina Bread with Biga and Sour

DESIRED DOUGH TEMPERATURE: 78°F/26°C

INGREDIENT	US	METRIC	BAKERS' PERCENTAGE
Biga			
Water	1 lb 4½ oz	585 g	11.7%
Durum flour	1 lb 3 oz	530 g	10.6%
Semolina flour	1 lb 3 oz	530 g	10.6%
Yeast, instant dry	⅛ tsp	0.5 g	—
Final Dough			
Water	6 lb 6 oz	2.9 kg	58.6%
Biga	3 lb 8½ oz	1.6 kg	33.0%
Durum Sour (page 204)	1 lb 6½ oz	640 g	12.8%
Olive oil	8 oz	220 g	4.4%
Malt syrup	1 oz	30 g	0.6%
Durum flour	2 lb 7 oz	1.1 kg	21.6%
Semolina flour	2 lb 7 oz	1.1 kg	21.6%
Bread flour	4 lb	1.8 kg	35.5%
Yeast, instant dry	1 oz	30 g	0.5%
Salt	4¼ oz	120 g	2.4%
Total Dough Weight	21 lb 1¼ oz	9.5 kg	191.0%
Semolina flour or sesame seeds	As needed	As needed	

1. Prepare the biga and durum sour 18 hours before mixing the final dough (see pages 125 and 204).
2. To prepare the final dough, combine the water, biga, durum sour, oil, malt syrup, durum flour, semolina flour, bread flour, and yeast in a mixer, and mix on low speed until homogenous, about 3 minutes. Cover the dough in the mixer bowl and rest for 10 to 15 minutes to autolyse.
3. Mix the dough on low speed for about 2 minutes, and add the salt. Increase the speed to high and mix until the dough reaches the improved stage of gluten development, about 2 minutes.
4. Remove the dough from the mixer and bulk-ferment, covered, for about 45 minutes. Fold and ferment for an additional 15 minutes.
5. Divide the dough into 1 lb 2 oz/510 g pieces and preshape into logs. Bench-rest the dough for 10 to 15 minutes.
6. Shape the dough into 18-in/46-cm baguettes and roll the tops in semolina flour or sesame seeds. Place on a couche. Proof the dough for 30 to 40 minutes or as needed, depending on the environment.
7. Score and load the dough into a 470°F/243°C oven and bake for 15 minutes, until lightly browned. Open the vent and bake until golden brown, about 8 minutes.
8. Place the finished breads on a rack to cool.

Semolina Bread with Golden Raisins and Fennel Seeds

DESIRED DOUGH TEMPERATURE: 78°F/26°C

INGREDIENT	US	METRIC	BAKERS' PERCENTAGE
Semolina Bread with Biga and Sour (page 184)	9 lb	4.1 kg	100.0%
Golden raisins, plumped	1 lb 13 oz	820 g	20.0%
Fennel seeds	2 oz	50 g	1.2%
Cornmeal	2 oz	50 g	1.2%
Total Dough Weight	11 lb 1 oz	5 kg	122.4%
Cornmeal	As needed	As needed	

1. Prepare the Semolina Bread according to the recipe on page 184.
2. Add the raisins, fennel seeds, and cornmeal. Mix on low speed until homogenous about 2 minutes.
3. Remove the dough from the mixer and bulk-ferment for 45 minutes. Fold and ferment an additional 15 minutes.
4. Divide the dough into 1 lb/450 g pieces and pre-shape into 10-in/25-cm oblong logs. Bench-rest the dough for 10 to 15 minutes.
5. Shape the dough into 18-in/46-cm baguettes and roll the tops in additional cornmeal. Place on a couche. Proof the dough for 30 to 40 minutes or as needed, depending on the environment.
6. Score straight down the middle of the dough. Load the dough into a 470°F/243°C oven and bake for 15 minutes until lightly browned. Open the vent and bake until golden brown, about 8 minutes.
7. Place the finished breads on a rack to cool.

Pane Siciliano

DESIRED DOUGH TEMPERATURE: 76°F/24°C

INGREDIENT	US	METRIC	BAKERS' PERCENTAGE
Biga, 65% hydration (page 125)	5 lb 1 oz	2.3 kg	45.0%
Water	7 lb 15 oz	3.6 kg	72.0%
Durum flour	11 lb	5 kg	100.0%
Yeast, instant dry	2 oz	55 g	1.0%
Salt	4½ oz	130 g	2.6%
Total Dough Weight	24 lb 6½ oz	11 kg	220.6%
Sesame seeds	As needed	As needed	
Semolina flour, for dusting	As needed	As needed	

Pane Siciliano dough should be smooth and yellow, because of its durum flour.

Roll the dough into tapered oblongs before moistening on a damp towel and rolling in sesame seeds.

1. Break the biga into smaller pieces and combine with the water and flour in a mixer. Mix until homogenous, about 2 minutes.
2. Stop the mixer and sprinkle the yeast on the top of the dough. Cover the dough in the mixer and rest for 10 to 15 minutes to autolyse.
3. Mix on low speed for 30 seconds, add the salt, and mix for an additional 30 seconds. Increase the speed to high and mix until the dough reaches the intense stage of gluten development, about 1 minute.
4. Remove the dough from the mixer and bulk-ferment, covered, for 1½ hours.
5. Divide the dough into 1 lb 2 oz/510 g pieces and preshape into tapered oblongs about 14 in/35 cm long.
6. Roll the dough onto a sheet lined with wet towels, then roll the dough pieces in the sesame seeds.
7. Shape the dough into crescents as they are placed on semolina-dusted boards. Proof for about 30 minutes or as needed, depending on the environment.
8. Score with a bench scraper, cutting ½ in/1 cm into the dough along the outside edge of the crescent. The marks should be about 1 in/2.5 cm apart.
9. Load the dough into a 450°F/232°C oven. Bake until golden brown, about 40 minutes.
10. Place the finished breads on a rack to cool.

Use a bench knife to score all the way through the outside edge of the shaped crescent to give the bread its characteristic shape.

Both the crescent and the "S" shape are traditional methods of shaping this dough.

Bagels

DESIRED DOUGH TEMPERATURE: 78°F/26°C

INGREDIENT	US	METRIC	BAKERS' PERCENTAGE
Sponge			
Water	3 lb 5 oz	1.5 kg	30.0%
Malt syrup	1 oz	30 g	0.5%
Bread flour	3 lb 5 oz	1.5 kg	30.0%
Yeast, instant dry	2½ oz	75 g	0.1%
Final Dough			
Water	3 lb 1½ oz	1.4 kg	27.5%
Sponge	6 lb 10 oz	3 kg	60.5%
Malt syrup	1½ oz	45 g	0.9%
Bread flour	7 lb 1 oz	3.2 kg	70.0%
Yeast, instant dry	¾ oz	20 g	0.4%
Salt	4 oz	115 g	2.3%
Total Dough Weight	15 lb 2¾ oz	7.8 kg	161.6%
Boiling Mixture			
Water	8 lb	3.6 kg	
Malt syrup	1 oz	30 g	
Toppings of choice (optional)	As needed	As needed	

1. To prepare the sponge, combine the water, malt syrup, flour, and yeast in a mixer. Mix on low speed for 4 minutes. Cover and ferment until doubled in size, about 1½ hours.

2. To prepare the final dough, combine the water, sponge, malt syrup, flour, yeast, and salt in the bowl of a mixer and mix for 4 minutes on low speed. Increase the speed and mix on high until the dough reaches the intense stage of gluten development, about 4 minutes.

3. Immediately begin dividing the dough. For large bagels, divide the dough into 5¼ oz/150 g pieces. If using a dough divider, scale 6 lb/2.7 kg press, which will yield 36 equal portions.

4. Preshape each piece of dough into a 3-in/7½-cm log. Bench-rest the dough covered, seam side down, for about 10 minutes.

5. Continue to roll each piece of dough until it is 18 in/46 cm to 20 in/51 cm long, with slightly tapered ends. Pinch one end to flatten and wrap it around to meet the other end, sealing it well with the palm of your hand.

6. In a large stockpot over moderate heat, bring the water and malt syrup to a boil.
7. Drop the bagels into the boiling mixture, a few at a time. They will fall to the bottom of the pot immediately before floating to the surface.
8. Once the bagels begin to float, allow 20 to 30 seconds of cooking time. Use a spider to transfer the bagels to a draining rack.
9. Immediately dip the bagels in any toppings, if using, and place on parchment-lined sheet trays.
10. Load the bagels into a 525°F/274°C oven. Immediately drop the temperature to 500°F/260°C and bake until the bottoms are golden brown, 10 to 14 minutes.
11. Place the finished bagels on a rack to cool.

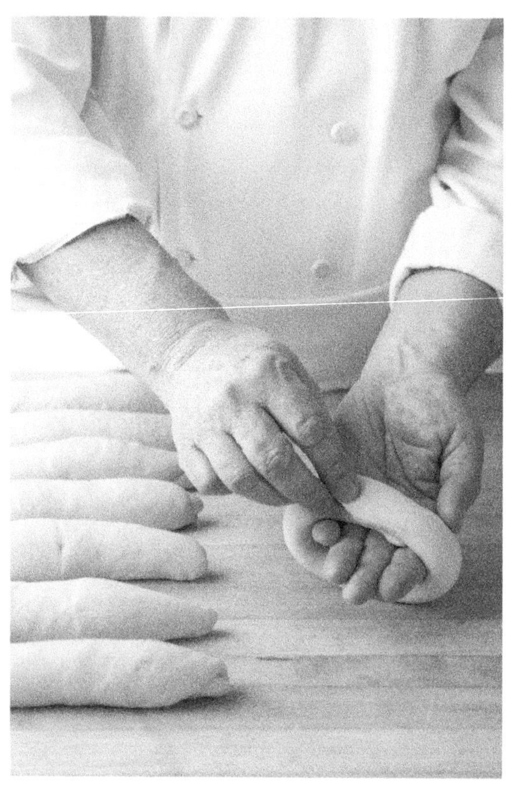

Wrap the dough so that the two ends overlap slightly, and press firmly to close the seam.

Remove the bagel from the boiling mixture when the exterior feels gelatinous. The dough will still be soft.

Dip the bagels in seeds or other toppings as soon as they are removed from the boiling mixture. The moisture will help adhere the garnish to the dough.

Finished bagels should be golden brown, with a visible seam to show that they were hand-shaped.

Blueberry Bagels

DESIRED DOUGH TEMPERATURE: 78°F/26°C

INGREDIENT	US	METRIC	BAKERS' PERCENTAGE
Sponge			
Water	3 lb 5 oz	1.5 kg	30.0%
Bread flour	3 lb 5 oz	1.5 kg	30.0%
Yeast, instant dry	¼ oz	5 g	0.1%
Malt syrup	1 oz	30 g	0.5%
Final Dough			
Sponge	6 lb 10 oz	3 kg	60.6%
Water	2 lb 14 oz	1 kg	27.0%
Bread flour	7 lb 7½ oz	3.4 kg	70.0%
Yeast, instant dry	½ oz	15 g	0.3%
Salt	3½ oz	100 g	2.0%
Malt syrup	1½ oz	35 g	0.8%
Dried blueberries, rehydrated	1 lb 6½ oz	745 g	13.2%
Total Dough Weight	16 lb 11½ oz	8.3 kg	173.9%
Boiling Mixture			
Water	8 lb	3.6 kg	
Malt syrup	1 oz	30 g	

1. To prepare the sponge, combine the water, flour, yeast, and malt syrup in a mixer and mix on low speed until homogenous, about 4 minutes. Ferment until doubled in size, 60 to 90 minutes.
2. To prepare the final dough, combine the sponge, water, flour, yeast, salt, and malt syrup in a mixer. Mix on low speed until homogenous, about 4 minutes.
3. Toss the blueberries with enough bread flour to lightly coat and add to the dough. Mix on low speed to incorporate, about 1 minute.
4. Divide the dough immediately. For large bagels, divide the dough into 5¼ oz/150 g pieces. For mini bagels, divide the dough into 2½ oz/75 g pieces.
5. Preshape each piece into a 3-in/8-cm log. Bench-rest the dough, seam side down, for about 10 minutes.
6. Continue to roll each piece of dough until 18 to 20 in/46 to 51 cm long. Pinch one end to flatten and wrap it around to meet the other end, sealing it well with the palm of your hand.
7. In a large stockpot over moderate heat, bring the water and malt syrup to a boil.
8. Drop the bagels into the boiling mixture, a few at a time. They will fall to the bottom of the pot immediately before floating to the surface.
9. Once the bagels begin to float, allow 20 to 30 seconds of cooking time. Use a spider to transfer the bagels to a draining rack. Place on parchment-lined sheet trays.
10. Load the bagels into a 525°F/274°C oven. Immediately drop the temperature to 500°F/260°C and bake until the bottoms are golden brown, 10 to 14 minutes.
11. Place the finished bagels on a rack to cool.

Cinnamon Raisin Bagels

DESIRED DOUGH TEMPERATURE: 78°F/26°C

INGREDIENT	US	METRIC	BAKERS' PERCENTAGE
Sponge			
Water	3 lb 5 oz	1.5 kg	30.0%
Malt syrup	1 oz	30 g	0.5%
High-gluten flour	3 lb 5 oz	1.5 kg	30.0%
Yeast, instant dry	¼ oz	5 g	0.1%
Final Dough			
Water	3 lb 1½ oz	1.4 kg	27.7%
Sponge	6 lb 10 oz	3 kg	60.5%
Malt syrup	1½ oz	45 g	0.9%
Bread flour	7 lb 11 oz	3.5 kg	70.0%
Yeast, instant dry	½ oz	15 g	0.3%
Salt	3½ oz	100 g	2.0%
Dark raisins	1 lb 10½ oz	750 g	15.0%
Ground cinnamon	¼ oz	10 g	0.6%
Total Dough Weight	19 lb 6¾ oz	8.8 kg	177.0%
Boiling Mixture			
Water	8 lb	3.6 kg	
Malt syrup	1 oz	30 g	
Toppings of choice (optional)	As needed	As needed	

1. To prepare the sponge, combine the water, malt syrup, flour, and yeast in a mixer. Mix on low speed for 4 minutes. Ferment until doubled in size, about 1½ hours.
2. To prepare the final dough, combine the water, sponge, malt syrup, flour, yeast, and salt in a mixer and mix for 4 minutes on low speed. Increase the speed and mix on high until the dough reaches the intense stage of gluten development, about 4 minutes.
3. Add the raisins and mix on low speed for about 1 minute. Add the cinnamon and mix until it is "swirled" into the dough, about 1 minute.
4. Immediately begin dividing the dough. For large bagels, divide the dough into 5¼ oz/150 g pieces. If using a dough divider, scale 6 lb/2.7 kg press, which will yield 36 equal portions.
5. Preshape each piece of dough into a 3-in/8-cm log. Bench-rest the dough, seam side down, for about 10 minutes.
6. Continue to roll each piece of dough until 18 to 20 in/46 to 51 cm long. Pinch one end to flatten and wrap it around to meet the other end, sealing it well with the palm of your hand.
7. In a large stockpot over moderate heat, bring the water and malt syrup to a boil.
8. Drop the bagels into the boiling mixture, a few at a time. They will fall to the bottom of the pot immediately before floating to the surface.
9. Once the bagels begin to float, allow 20 to 30 seconds of cooking time. Use a spider to transfer the bagels to a draining rack.
10. Immediately dip the bagels in toppings, if using, and place on parchment-lined sheet trays.
11. Load the bagels into a 525°F/274°C oven. Immediately drop the temperature to 500°F/260°C and bake until the bottoms are golden brown, 10 to 14 minutes.
12. Place the finished bagels on a rack to cool.

Sugar Crunch Bagels

DESIRED DOUGH TEMPERATURE: 78°F/26°C

INGREDIENT	US	METRIC	BAKERS' PERCENTAGE
Sponge			
Water	3 lb 5 oz	1.5 kg	30.0%
Bread flour	3 lb 5 oz	1.5 kg	30.0%
Yeast, instant dry	¼ oz	5 g	0.1%
Malt syrup	1 oz	30 g	0.5%
Final Dough			
Water	3 lb 1½ oz	1.4 kg	27.0%
Bread flour	7 lb 11 oz	3.5 kg	70.0%
Yeast, instant dry	¾ oz	20 g	0.4%
Salt	4 oz	115 g	2.3%
Malt syrup	1½ oz	45 g	0.9%
Sponge	6 lb 10 oz	3.0 kg	60.6%
Pearl sugar	1 lb 1¾ oz	500 g	10.0%
Total Dough Weight	18 lb 14½ oz	8.6 kg	171.2%
Boiling Mixture			
Water	8 lb	3.6 kg	
Malt syrup	1 oz	30 g	

1. Prepare the sponge the same day as the final dough. Combine the water, flour, yeast, and malt syrup in a mixer, and mix on low speed until homogenous. Ferment until doubled in size, about 1½ hours.
2. To prepare the final dough, combine the water, flour, yeast, salt, malt syrup, and sponge in a mixer, and mix on low speed until homogenous. Increase the speed to high and mix until the dough reaches the intense stage of gluten development, about 6 minutes.
3. Add the pearl sugar and mix just until combined, about 1 minute.
4. Immediately divide the dough into 5¼ oz/150 g pieces and preshape each piece into a 3-in/8-cm log. Bench-rest the dough, seam side down, for about 10 minutes.
5. Continue to roll each piece of dough until 8 to 20 in/46 to 51 cm long. Pinch one end to flatten and wrap it around to meet the other end, sealing it well with the palm of your hand.
6. In a large stockpot over moderate heat, bring the water and malt syrup to a boil.
7. Drop the bagels into the boiling mixture, a few at a time. They will fall to the bottom of the pot immediately, before floating to the surface.
8. Once the bagels begin to float, allow 20 to 30 seconds of cooking time. Use a spider to transfer the bagels to a draining rack.
9. Place the bagels on a parchment-lined sheet tray.
10. Load the bagels into a 525°F/274°C oven. Immediately drop the temperature to 500°F/260°C and bake until the bottoms are golden brown, 10 to 14 minutes.
11. Place the finished bagels on a rack to cool.

Sun-dried Tomato and Basil Bagels

DESIRED DOUGH TEMPERATURE: 75°F/24°C

INGREDIENT	US	METRIC	BAKERS' PERCENTAGE
Sponge			
Water	3 lb 5 oz	1.5 kg	30.0%
Bread flour	3 lb 5 oz	1.5 kg	30.0%
Yeast, instant dry	¼ oz	5 g	0.1%
Malt syrup	1 oz	30 g	0.5%
Final Dough			
Sponge	6 lb 10 oz	3 kg	60.6%
Water	3 lb 1½ oz	1.4 kg	27.0%
Bread flour	7 lb 11 oz	3.5 kg	70.0%
Yeast, instant dry	½ oz	15 g	0.3%
Salt	3½ oz	100 g	2.0%
Malt syrup	1½ oz	40 g	0.8%
Sun-dried tomatoes, chopped	1 lb 7¼ oz	660 g	13.2%
Basil, cut in chiffonade	2½ oz	75 g	1.5%
Bread flour	5½ oz	160 g	3.2%
Total Dough Weight	19 lb 11¼ oz	9 kg	178.7%
Boiling Mixture			
Water	8 lb	3.6 kg	
Malt syrup	1 oz	30 g	

1. To prepare the sponge, combine the water, flour, yeast, and malt syrup in a mixer bowl and mix on low speed until homogenous, about 4 minutes. Ferment until doubled in size, 60 to 90 minutes.
2. To prepare the final dough, combine the sponge, water, flour, yeast, salt, and malt syrup in a mixer bowl. Mix on low speed until homogenous, about 4 minutes.
3. Increase the speed to high and mix until the dough reaches the intense stage of gluten development, about 6 minutes.
4. Add the tomatoes and basil, and mix on low speed until incorporated, about 1 minute.
5. Immediately divide the dough. For large bagels, divide the dough into 5¼-oz/150-g pieces and preshape each piece into a 3-in/7½-cm log. For mini bagels, divide the dough into 2½-oz/75-g pieces. Bench-rest, seam side down, for about 10 minutes.
6. Continue to roll each piece of dough until 18 to 20 in/46 to 51 cm long. Pinch one end to flatten and wrap it around to meet the other end, sealing it well with the palm of your hand.
7. In a large stockpot over moderate heat, bring the water and malt syrup to a boil.
8. Drop the bagels into the boiling mixture, a few at a time. They will fall to the bottom of the pot immediately before floating to the surface.
9. Once the bagels begin to float, allow 20 to 30 seconds of cooking time. Use a spider to transfer the bagels to a draining rack. Place on parchment-lined sheet trays.
10. Load the bagels into a 525°F/274°C oven. Immediately drop the temperature to 500°F/260°C and bake until the bottoms are golden brown, 10 to 14 minutes.
11. Place the finished bagels on a rack to cool.

Bialys

DESIRED DOUGH TEMPERATURE: 78°F/26°C

INGREDIENT	US	METRIC	BAKERS' PERCENTAGE
Sponge			
Water	3 lb 5 oz	1.5 kg	30.0%
Bread flour	3 lb 5 oz	1.5 kg	30.0%
Yeast, instant dry	¼ oz	5 g	0.1%
Final Dough			
Water	3 lb 8½ oz	1.6 kg	31.0%
Malt syrup	¾ oz	20 g	0.4%
Sponge	6 lb	2.7 kg	60.0%
Bread flour	7 lb 1 oz	3.2 kg	70.0%
Yeast, instant dry	¾ oz	20 g	0.4%
Salt	4 oz	115 g	2.3%
Total Dough Weight	16 lb 15 oz	7.7 kg	164.3%
Filling			
Onions, diced and sautéed	3 lb 1½ oz	1.4 kg	
Poppy seeds	8 oz	230 g	
Olive oil	2 oz	60 g	

1. Prepare the sponge the same day as the final dough. Combine the water, flour, and yeast in a mixer and mix on low speed until homogenous, about 4 minutes. Place in a container and ferment until doubled in size, about 1½ hours.
2. To prepare the final dough, combine the water, malt syrup, sponge, flour, yeast, and salt in a mixer. Mix on low speed until homogenous, about 4 minutes. Increase the speed to high and mix until the dough reaches the intense stage of gluten development, about 6 minutes.
3. Immediately divide the dough into 4½ oz/130 g pieces (or a 5 lb/2.30 kg press to yield 18 pieces on a dough divider) upon removing the dough from mixer. Preshape into rounds and bench-rest the dough for about 15 minutes.
4. Shape the dough by pressing a dowel into the center of the round. Do not push the dowel all the way through the dough. Stretch the bialy to create an open center that is about 1 in/2.5 cm. Transfer to a parchment-lined sheet tray.
5. Prepare the filling by combining the onions, poppy seeds, and oil. Spoon about 1 oz/30 g filling into the center of each bialy.
6. Load the bialys into a 475°F/246°C convection oven, steam, and immediately lower the temperature to 425°F/218°C. Bake until golden brown, about 12 minutes.
7. Place the finished bialys on a rack to cool.

Everything Cream-Cheese Spread

YIELD: 11 LB/5 KG

INGREDIENT	US	METRIC
Cream cheese	10 lb 5¾ oz	4.7 kg
Garlic, minced	1 oz	30 g
Onion, minced	8 oz	230 g
Poppy seeds	2 oz	50 g
Sesame seeds	2¾ oz	75 g
Red pepper flakes	1 oz	30 g
Salt	As needed	As needed
Pepper	As needed	As needed

1. Place the cream cheese in a mixer and mix with the paddle attachment to soften. Add the garlic, onions, poppy seeds, sesame seeds, and red pepper flakes. Mix to combine.
2. Season with salt and pepper.
3. Cover and refrigerate until ready to use.

Sun-dried Tomato Cream-Cheese Spread

YIELD: 3 LB 12.8 OZ/1.7 KG

INGREDIENT	US	METRIC
Cream cheese	3 lb 1½ oz	1.4 kg
Garlic, minced	¾ oz	20 g
Sun-dried tomatoes, puréed	12 oz	340 g
Basil, cut in chiffonade	¾ oz	20 g
Salt	As needed	As needed
Pepper	As needed	As needed

1. Soften the cream cheese in a mixer with the paddle attachment. Add the garlic, tomato purée, and basil. Season with salt and pepper.
2. Mix until well combined, making sure to scrape the bowl often, about 1 minute.
3. Cover and refrigerate until ready to use.

Vegetable Cream-Cheese Spread

YIELD: 4 LB 3 OZ/1.9 KG

INGREDIENT	US	METRIC
Cream cheese	3 lb	1.4 kg
Garlic, minced	¾ oz	20 g
Green onions, sliced	3½ oz	105 g
Carrots, grated	6¾ oz	190 g
Green pepper, diced	7½ oz	210 g
Salt	As needed	As needed
Pepper	As needed	As needed

1. Soften the cream cheese in a mixer, using the paddle attachment. Add the garlic, green onions, carrots, and peppers and mix to combine.
2. Season with salt and pepper.
3. Cover and refrigerate until ready to use.

Biga

INGREDIENT	US	METRIC	BAKERS' PERCENTAGE
Water, 55–65°F/13–18°C (see Note)	2 lb	915 g	57.4%
Bread flour	3 lb 12 oz	1.7 kg	100.0%
Yeast, instant dry	¼ tsp	1 g	—
Total Dough Weight	6 lb	2.7 kg	157.4%

1. Combine the water, flour, and yeast in a mixer, and mix on low speed until homogenous, about 2 minutes.
2. Place the dough into a container large enough to allow it to double in size. Store, covered, at room temperature.

CHEF'S NOTE

This is a dry preferment. There is enough water to saturate the flour, but it will take a few minutes to incorporate all the flour into the dough.

If a formula calls for a biga of a certain hydration level (for example, Biga, 85% hydration), adjust the water in this formula so that it is 85% of the biga's flour weight.

Pâte Fermentée

INGREDIENT	US	METRIC	BAKERS' PERCENTAGE
Water, 55–65°F/13–18°C	7 lb 4½ oz	3.3 kg	66.7%
Bread flour	11 lb	5 kg	100.0%
Salt	4 oz	110 g	2.2%
Yeast, instant dry	½ oz	15 g	0.3%
Malt syrup	1 oz	30 g	0.5%
Total Dough Weight	18 lb 10 oz	8.5 kg	169.7%

1. Combine the water, flour, salt, yeast, and malt syrup in a mixer and mix until homogenous, about 2 minutes.
2. Place the dough in a lightly oiled container large enough to allow it to double in size. Cover.
3. The pâte fermentée can be left at room temperature to use the next day. If not using the next day, leave it out for 1 hour and then refrigerate for up to 3 days.

Sour Starter Development

INGREDIENT	US	METRIC	BAKERS' PERCENTAGE
Day 1			
Water, room temperature	4 oz	115 g	100.0%
Bread flour	4 oz	115 g	100.0%
Day 2			
Water, room temperature	4 oz	115 g	100.0%
Bread flour	4 oz	115 g	100.0%
Sour starter (from Day 1)	8 oz	230 g	200.0 %
Day 3			
Water, room temperature	4 oz	115 g	100.0%
Bread flour	4 oz	115 g	100.0%
Sour starter (from Day 2)	4 oz	115 g	100.0%
Day 4			
Water, room temperature	4 oz	115 g	100.0%
Bread flour	4 oz	115 g	100.0%
Sour starter (from Day 3)	4 oz	115 g	100.0%
Day 5			
Water, room temperature	8 oz	225 g	66.7%
Bread flour	12 oz	340 g	100.0%
Sour starter (from Day 4)	4 oz	115 g	33.3%

Day 1. Using organic or unbleached flour (see Note), combine the flour and water in a container and stir until homogenous. Leave uncovered at room temperature.

Day 2. Combine the flour, water, and the entire sour starter from Day 1 in a container. Stir until homogenous. Leave uncovered at room temperature.

Day 3. Combine the flour, water, and half the previous day's starter (discard the remainder). Stir until homogenous. Leave covered at room temperature.

Day 4. Combine the flour, water, and half the previous day's starter. Stir until homogenous. Leave covered at room temperature.

Day 5. Combine the flour, water, and half the previous day's starter. Stir until homogenous. Leave covered at room temperature.

CHEF'S NOTE

To keep the sour starter alive for use, continue to feed it using the formula for Day 5.

To preserve the sour without feeding it daily, give the sour a Day 5 feeding and place it immediately into the refrigerator. It can stay in the refrigerator for 3 weeks.

To refresh your sour, take it out of the refrigerator for one day and then start the Day 5 feeding for 2 to 3 days. The sour can then be used as normal.

It is possible to use up to 25 percent of an alternative flour (durum, rye, or whole wheat). If you choose to use an alternative flour, keep the flour percentages the same throughout the feedings and in the final dough.

Although this starter takes only 5 days, it is best to feed your starter for about 14 days before using. This allows the yeast and bacteria in the dough to fully develop and become healthy.

White Liquid Sour Feeding

INGREDIENT	US	METRIC	BAKERS' PERCENTAGE
White Sourdough Starter (page 142)	10¾ oz	305 g	13.0%
Water, 55–65°/13–18°C	5 lb 4½ oz	2.4 kg	100.0%
Bread flour	5 lb 4½ oz	2.4 kg	100.0%
Total Dough Weight	11 lb 3¾ oz	5.1 kg	213.0%

1. Mix the feeding 18 hours before using it. In a large bowl or plastic container, dissolve the white sourdough starter in the water.
2. Add the flour and mix by hand until the dough is homogenous.
3. Place in a container large enough for dough to double in size. Cover and leave at room temperature to ferment for 10 hours.
4. Feed again and ferment for an additional 8 to 10 hours.

White Sour Feeding

INGREDIENT	US	METRIC	BAKERS' PERCENTAGE
Water, 55–65°/13–18°C	3 lb 8½ oz	1.6 kg	62.5%
Bread flour	5 lb 1 oz	2.3 kg	90.0%
Whole wheat flour	9 oz	260 g	10.0%
White Sourdough Starter (page 142)	1 lb 11½ oz	780 g	30.0%
Total Dough Weight	10 lb 14 oz	5 kg	192.5%

1. Mix the feeding 18 hours before using. Combine the water, bread flour, whole wheat flour, and white sourdough starter in a small mixer and mix on low speed until homogenous, about 3 minutes.
2. Place the dough in a container large enough to allow it to double in size. Cover and leave at room temperature to ferment for 18 hours.

Semolina Biga

INGREDIENT	US	METRIC	BAKERS' PERCENTAGE
Water, 55–65°F/13–18°C	1 lb 15 oz	885 g	54.8%
Durum flour	1 lb 12½ oz	805 g	50.0%
Semolina flour	1 lb 12½ oz	805 g	50.0%
Yeast, instant dry	⅛ tsp	0.5 g	—
Total Dough Weight	5 lb 8 oz	2.5 kg	154.8%

1. Combine the water, durum flour, semolina flour, and yeast in a mixer. Mix on low speed until homogenous, about 2 minutes.
2. Transfer the dough to a container big enough for it to double in size. Ferment, covered, at room temperature.

CHEF'S NOTE

This biga works well with Semolina Bread (page 182) or with any other breads containing semolina flour, the flour typically found in Italian breads.

Semolina Sour Feeding

INGREDIENT	US	METRIC	BAKERS' PERCENTAGE
Water, 60°F/16°C	3 lb 12 oz	1.7 kg	70.0%
White Sourdough Starter (page 142)	1 lb 13 oz	820 g	33.3%
Bread flour	3 lb 5 oz	1.5 kg	60.0%
Semolina flour	1 lb 1¼ oz	490 g	20.0%
Durum flour	1 lb 1¼ oz	490 g	20.0%
Total Dough Weight	11 lb	5 kg	203.3%

1. Mix the feeding 18 hours prior to using. Combine the water and white sourdough starter in a mixer and mix with a dough hook on low speed until combined, about 1 minute.
2. Add the bread flour, semolina flour, and durum flour, and mix on low speed until homogenous, about 3 minutes.
3. Transfer to a container large enough to allow the dough to double in size. Cover and ferment at room temperature for 18 hours.

Durum and Whole Wheat Sour Feeding

INGREDIENT	US	METRIC	BAKERS' PERCENTAGE
Water, 55–65°F/13–18°C	3 lb 12 oz	1.7 kg	70.0%
White Sourdough Starter (page 142)	1 lb 13 oz	820 g	33.3%
Bread flour	3 lb 5 oz	1.5 kg	60.0%
Durum flour	1 lb 1¼ oz	490 g	20.0%
Whole wheat flour	1 lb 1¼ oz	490 g	20.0%
Total Dough Weight	11 lb	5 kg	203.3%

1. Mix the feeding 18 hours before using. Combine the water and white sourdough starter in a mixer fitted with the dough hook. Mix on low speed until combined, about 1 minute.
2. Add the bread flour, durum flour, and whole wheat flour, and mix on low speed until the ingredients are homogenous, about 3 minutes.
3. Place in a container that will allow the dough to double in size. Ferment at room temperature for at least 18 hours.

ENRICHED BREAD AND ROLLS

6

The term "enriched dough" describes dough that is made using fats, sweeteners, dairy, and/or eggs. There are two basic categories of enriched doughs: those in which the dough itself is enriched with the addition of fats or sweeteners, and those that are laminated with fat. These two categories can be combined to create laminated enriched doughs, such as a laminated brioche.

Generally speaking, enriched breads are soft and chewy, such as brioche. Some enriched doughs are handled in a specific way to give them an alternative texture, such as croissant and Danish doughs, which are laminated. Other enriched doughs gain their flavor from inclusions, fillings, or toppings, such as cinnamon rolls. Enriched doughs can be made with a range of fat amounts, up to equal amounts of fat and flour, depending on the desired outcome.

Traditional enriched breads, such as stollen, have standard garnishes and inclusion; in this case, they are typically raisins, orange peel, and lemon peel. However, bakers are now making stollen with other garnishes as well. It is up to bakers to determine what the ideal product is for their operations.

SPONGE

A sponge is a type of preferment that originated in Europe, commonly employed for the production of a sweet dough (basic enriched dough). The process for making a sponge is similar to the process for making a poolish (see page 124). The primary difference is in the hydration levels, as the hydration for a sponge is determined by the consistency of the final dough. Additionally, a sponge should be used as soon as it reaches full maturation. When many bubbles are evident and the sponge appears to crack, it has matured and should immediately be added to a finished dough. A sponge is generally added at the beginning of the mixing process.

Ratio: 100% flour, 60 to 65% water, 0.2 to 1.5% yeast

Fermentation Time: From 15 minutes to 1 hour at room temperature. A sponge that has not matured will lack the proper yeast development to be a beneficial preferment,

Enriched breads. From the left: Gugelhopf, Brioche, Brioche à Tête, Pan de Muerto, Panettone, Danish ring, Challah, Babka.

while an overmature sponge could negatively affect the strength of the dough, owing to excessive alcohol. Sponges should not be fermented for excessive times or overnight, as done with some other preferments.

Use: Easily worked into the dough—a sponge is slightly stiffer than the final dough, so it is easier to handle. Sponges are most commonly used for sweet doughs, though also for bagels and bialys (see Chapter 5). A stiff consistency means increased strength, which can be beneficial for the final bread dough, especially for highly enriched doughs, which can suffer from a weakening of the gluten bonds.

LAMINATION

Laminated doughs are created when butter and dough are layered through a series of folds; the optimal result is even layers of fat and dough. The fat separates the dough layers from each other. Steam is released from the water present in the dough and butter, which separates the layers during baking. The starch in the dough gelatinizes and stabilizes during baking, allowing the final dough to maintain its structure. This process is commonly done with pastries (known as puff pastry), but also with yeast-raised doughs (croissant and Danish, which combine physical and biological leavening).

The process is not a simple one, though, and precise layering techniques obtain the proper end product. For example, folding the dough too many times can cause the fat and dough to merge, while folding too few times can prevent the dough from properly rising when baked. There are two folds used when making puff pastry: a three-fold, and a four-fold. These folds are used distinctly or in combination to create the many signature layers of a laminated dough.

Three-Fold

To create a three-fold, the dough is rolled out to a rectangle of the proper thickness. The sheet of pastry is divided visually into thirds, and one of the outer thirds is folded over the middle third. The remaining outer third is then folded over the folded dough, as if folding a business letter. This three-fold procedure triples the number of layers in the dough.

Four-Fold

To create a four-fold, the dough is rolled out to a rectangle of the proper thickness. The sheet of pastry is divided visually into quarters, and the outer quarters are folded into the middle so that their edges meet. Then the dough is folded over as if closing a book. This four-fold quadruples the number of layers in the dough.

Steps of Lamination

When the dough has been mixed, the lamination can begin. The steps are as follows:

1. Shape the dough into a rough square or rectangle. Transfer to a sheet pan lined with parchment paper, wrap the dough in plastic wrap, and allow it to relax under refrigeration for 30 to 60 minutes.

2. While the dough is resting, prepare the butter block. The butter is worked either by hand or carefully with a mixer until it is smooth and malleable but not overly soft. Adding a small amount of flour to the butter may make it easier to work with and to absorb excess moisture in the butter. Mix the butter with the flour (if using) until smooth.

3. Transfer the butter to a sheet of parchment paper. Cover with another sheet of parchment and roll the butter into a rectangle. The butter block should be slightly less than half the size of the dough rectangle. Square off the edges, cover with plastic wrap, and

refrigerate until firm but still pliable. Do not allow the butter block to become too warm or too cold—the temperature of both is crucial to the success of the lamination, and they should be the same consistency. Butter that is too cold will crack and not roll smoothly into the dough. Butter that is too hot will melt and escape from the dough, preventing the layers from forming.

4. Roll out the dough to a square that is twice the size of the block of butter. Place the block of butter on top of the square of dough, so that each of the corners of the butter are pointing directly to the center of a side of the dough. Fold the dough edges over the butter block.
5. The first fold performed is a three-fold (see page 208). Complete the three-fold, then cover the dough in plastic wrap and allow it to rest for 30 minutes under refrigeration.
6. Turn the dough 90 degrees from its position before it was refrigerated and roll it out into a rectangle, making sure the edges are straight and the corners are again squared. Complete a four-fold (see page 208). Cover the dough with plastic wrap and allow it to rest for 30 minutes in the refrigerator.
7. Turn the dough 90 degrees from its position before it was refrigerated and roll it out into a rectangle, making sure the edges are straight and the corners are again squared. Complete another three-fold. Allow the dough to rest for an additional 30 minutes before rolling it out and shaping.

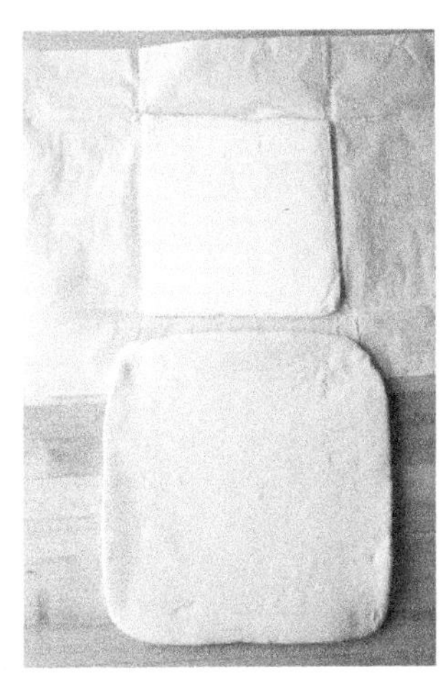

The roll-in butter should be about half the size of the rolled dough. Use folded parchment paper as a guide.

Lock in the butter by placing it in the center of the dough, with the corners of the butter facing the flat edges of the dough. Fold the dough in to meet in the butter center.

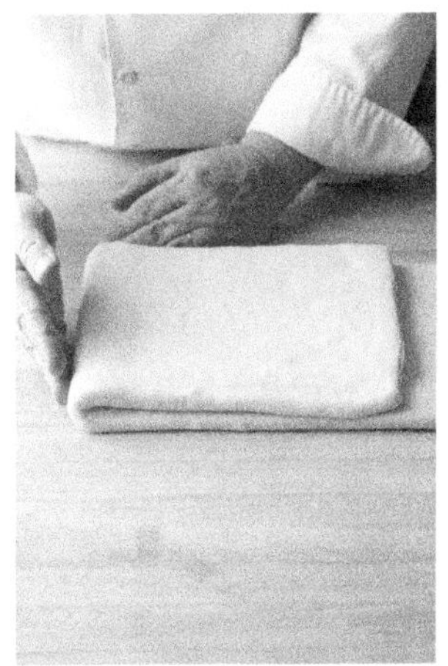

To give the dough a three-fold (also known as a letter fold), roll the dough into a rectangle and fold the right edge to two-thirds of the length of the rectangle.

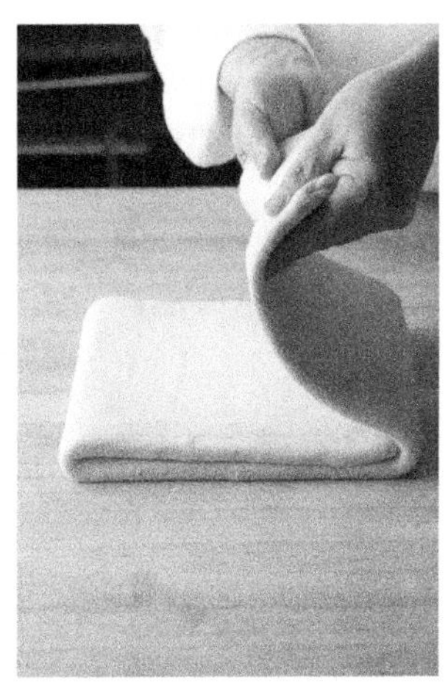

Fold the left edge of the dough over the first fold, like folding a letter.

Apple Cider Bread

DESIRED DOUGH TEMPERATURE: 78°F/26°C

INGREDIENT	US	METRIC	BAKERS' PERCENTAGE
Pâte Fermentée (page 199)	9 lb 4 oz	4.2 kg	83.4%
Sour cream	2 lb 2 oz	965 g	18.9%
Apple purée	1 lb 1¾ oz	500 g	15.1%
Apple cider	5 lb 8 oz	2.5 kg	49.2%
Malt syrup	1 oz	30 g	0.6%
Whole wheat flour	3 lb 1½ oz	1.4 kg	33.3%
Bread flour	7 lb 8 oz	3.4 kg	66.7%
Yeast, instant dry	1½ oz	40 g	0.8%
Salt	4 oz	115 g	2.3%
Total Dough Weight	29 lb	13.15 kg	270.3%

1. Combine all of the ingredients in a mixer. Mix on low speed for about 4 minutes. Increase the speed and mix on high until the dough reaches the improved stage of gluten development, about 3 minutes.
2. Remove the dough from the mixer and bulk-ferment, covered, until doubled in size, about 1 hour. Fold once halfway through the fermentation.
3. Divide the dough into 1 lb 5 oz/600 g pieces and preshape into rounds. Bench-rest the dough for 15 to 20 minutes.
4. Shape the dough into 9-in/23-cm bâtards. Proof for about 45 minutes, or as needed, depending on the environment.
5. Load the dough into a 470°F/243°C oven. Immediately steam and lower the temperature to 430°F/221°C. Bake until golden brown, 40 to 45 minutes.
6. Place the finished breads on a rack to cool.

Beer Bread

DESIRED DOUGH TEMPERATURE: 77°F/25°C

INGREDIENT	US	METRIC	BAKERS' PERCENTAGE
Pâte Fermentée (page 199)	3 lb 2 oz	1.4 kg	28.5%
Medium rye flour	1 lb 9¼ oz	720 g	14.4%
Bread flour	9 lb 8 oz	4.3 kg	85.6%
Yeast, instant dry	1½ oz	40 g	0.8%
Malt syrup	1 oz	30 g	0.5%
Cottage cheese	1 lb 9 oz	700 g	14.2%
Salt	4¼ oz	125 g	2.5%
Beer	5 lb 12 oz	2.6 kg	51.8%
Total Dough Weight	21 lb 15 oz	9.9 kg	198.3 %

1. Prepare the pâte fermentée 18 hours before mixing the final dough.
2. Combine all of the ingredients in a mixer, and mix on low speed for 4 minutes. Increase the speed and mix on high speed until the dough reaches the improved stage of gluten development, about 4 minutes.
3. Remove the dough from the mixer and bulk-ferment, covered, until doubled in size, about 1 hour. Fold once halfway through the fermentation.
4. Divide the dough into 1 lb 8 oz/680 g pieces and preshape into rounds. Bench-rest the dough for 15 minutes.
5. Shape the dough into 10-in/25-cm bâtards or boules. Place the bread onto floured couches or brotforms. Proof for about 45 minutes, or as needed, depending on the environment.
6. Load the dough into a 480°F/249°C oven. Immediately lower the temperature to 450°F/232°C and steam. Bake until golden brown, 40 to 45 minutes.
7. Place the finished breads on a rack to cool.

Sprouted Grain Bread

DESIRED DOUGH TEMPERATURE: 78°F/26°C

INGREDIENT	US	METRIC	BAKERS' PERCENTAGE
Biga			
Water	1 lb 9¼ oz	715 g	14.3%
Whole wheat flour	1 lb 14¼ oz	855 g	17.1%
Bread flour	13 oz	370 g	7.4%
Yeast, instant dry	¼ tsp	1 g	—
Final Dough			
Sprouted Grain Sponge (page 287)	11 lb	5 kg	100.5%
Biga	2 lb 3¼ oz	1.0 kg	38.8%
Water	5 lb 1 oz	2.3 kg	45.0%
Honey	2 lb 7 oz	1.1 kg	22.6%
Oil	3 lb 8½ oz	1.6 kg	31.5%
Malt syrup	1 oz	30 g	0.5%
Whole wheat flour	3 lb 8½ oz	1.6 kg	32.9%
Bread flour	4 lb 10 oz	2.1 kg	42.6%
Yeast, instant dry	2 oz	60 g	1.2%
Salt	8 oz	225 g	4.5%
Rolled oats, roasted	1 lb 12 oz	805 g	16.1%
Total Dough Weight	34 lb 13¼ oz	15.82 kg	336.2%
Rolled oats	As needed	As needed	

1. Prepare the biga (see page 125) 18 hours before mixing the final dough.
2. To prepare the final dough, combine the sponge, biga, water, honey, oil, and malt syrup in a mixer. Mix on low speed until homogenous, about 1 minute. Increase the speed to high and mix for an additional 1 minute to break up the sponge.
3. Add the whole wheat flour, bread flour, and yeast and mix on low speed until homogenous. Stop the mixer and rest the dough to autolyse for 15 to 20 minutes.
4. Add the salt and oats and resume mixing on low speed to combine, about 1 minute. Increase the speed to high and mix until the dough reaches the improved stage of gluten development, about 1 minute.

5. Remove the dough from the mixer and bulk-ferment, covered, for about 45 minutes. Fold the dough and ferment for an additional 15 minutes.

6. Divide the dough into 1 lb 12 oz/795 g pieces and preshape into rounds. Bench-rest the dough for 10 to 15 minutes.

7. Shape the dough into 10-in/25-cm oblong loaves. Roll in oats and place seam side down into oiled strap pans. Proof for 45 minutes to 1 hour, or as needed, depending on the environment.

8. Score the dough and load into a 425°F/218°C oven and immediately lower the temperature to 375°F/191°C. Steam and bake until golden brown, about 35 minutes.

9. Remove from the pans immediately after baking. Place the finished breads on a rack to cool.

Swedish Limpa

DESIRED DOUGH TEMPERATURE: 78°F/26°C

INGREDIENT	US	METRIC	BAKERS' PERCENTAGE
Rye Sour			
Water	1 lb 3¾ oz	560 g	70.0%
Medium whole rye flour, organic	1 lb 12 oz	800 g	100.0%
Basic Rye Sour (page 147)	1 lb 12 oz	800 g	100.0%
Final Dough			
Water	4 lb 13½ oz	2.2 kg	55.0%
High-gluten flour (13–15% protein)	5 lb 8 oz	2.5 kg	62.5%
Medium whole rye flour, organic	3 lb 5 oz	1.5 kg	37.5%
Molasses	1 lb 1¾ oz	500 g	12.5%
Sea salt	2 oz	60 g	1.5%
Fennel seeds, roasted and ground	1 oz	30 g	0.8%
Coriander seeds, roasted and ground	1 oz	30 g	0.8%
Orange zest, grated	3 oz	90 g	2.3%
Dried currants	1 lb 10½ oz	750 g	18.8%
Rye Sour	4 lb 6½ oz	2 kg	50.0%
Total Dough Weight	21 lb 4¼ oz	9.7 kg	241.7%

1. Make the rye sour 18 hours in advance with a desired dough temperature of 50°F/10°C.
2. To make the final dough, combine the water, high-gluten flour, rye flour, molasses, salt, and rye sour in a mixer, and mix on low speed for 4 to 5 minutes, until the dough comes together.
3. Add the fennel, coriander, orange zest, and currants and mix on low speed until all ingredients are combined and the dough is formed. Be careful not to overmix this dough.
4. Remove the dough from mixer and place in an oiled tub. Bulk-ferment for 1 hour, fold, ferment another hour, fold, and ferment a final 30 minutes.
5. Divide the dough into 2 lb/910 g pieces and preshape into a round. Bench-rest the dough for 10 to 15 minutes, covered.
6. Shape the dough into 18-in/46-cm bâtards and place into a lightly floured couche. Proof for 1½ hours, or as needed, depending on the environment.
7. Preheat a hearth oven to 450°F/232°C. Score and load the breads, steaming 1 second before loading and for 3 seconds after loading.
8. Lower the temperature to 425°F/218°C and bake for 30 minutes, opening the vent approximately 5 minutes into the baking.
9. Place the finished breads on a rack to cool.

Durum and Rosemary Rolls

DESIRED DOUGH TEMPERATURE: 78°F/26°C

INGREDIENT	US	METRIC	BAKERS' PERCENTAGE
Water	7 lb 15 oz	3.6 kg	72.0%
Bread flour	2 lb 14 oz	1.3 kg	25.9%
Durum flour	8 lb 2½ oz	3.7 kg	74.1%
Yeast, instant dry	1¾ oz	50 g	1.0%
Salt	4¼ oz	120 g	2.3%
Olive oil	5¼ oz	150 g	3.0%
Fresh rosemary, coarsely chopped	1¾ oz	50 g	1.0%
Total Dough Weight	19 lb 12½ oz	9 kg	179.3%

1. Combine the water, bread flour, durum flour, yeast, salt, and oil in a mixer and mix until homogenous, about 4 minutes.
2. Increase the speed to high and mix until the dough reaches the improved stage of gluten development, about 3 minutes.
3. Add the rosemary and mix until incorporated, about 1 minute.
4. Remove the dough from the mixer and bulk-ferment, covered, for about 1 hour. Fold and ferment for an additional 15 minutes.
5. Divide the dough into 4 lb/1.8 kg pieces and preshape into round presses. Bench-rest the dough for about 15 minutes.
6. Use a dough divider to divide each press into 36 pieces. For a more rustic appearance, do not machine-round the rolls. Place on a floured couche, about 5 rolls per row, and dust with durum flour. Proof the dough for 30 to 45 minutes, or as needed, depending on the environment.
7. Score the rolls once on top and load into a 475°F/246°C. Immediately lower the temperature to 425°F/218°C. Bake until golden brown, about 20 minutes.
8. Place the finished breads on a rack to cool.

Babka

DESIRED DOUGH TEMPERATURE: 75°F/24°C

INGREDIENT	US	METRIC	BAKERS' PERCENTAGE
Butter, room temperature	2 lb 3¼ oz	1 kg	19.8%
Bread flour	11 lb	5 kg	100.0%
Milk	4 lb	1.8 kg	35.3%
Eggs	1 lb 13½ oz	835 g	16.7%
Salt	2½ oz	75 g	1.5%
Lemon zest, grated	1 oz	30 g	0.5%
Ground mace	2½ tsp	5 g	0.1%
Yeast, instant dry	3¾ oz	110 g	2.2%
Brown sugar	1 lb 13½ oz	835 g	16.7%
Total Dough Weight	21 lb 5¾ oz	9.7 kg	192.8%
Poppy Seed Filling (page 218)	3 lb 5 oz	1.5 kg	
Apricot glaze	As needed	As needed	
Confectioners' sugar	As needed	As needed	
Streusel	As needed	As needed	

1. Divide the butter into 3 equal portions and set aside.
2. Combine the flour, milk, eggs, salt, lemon zest, mace, yeast, and one third of the butter in a mixer. Mix on low speed until homogenous, about 2 minutes.
3. Increase the speed and mix on high speed until the dough has reached the intense stage of gluten development, about 2 minutes.
4. Return the speed to low and blitz in the second third of butter, and then stream in the brown sugar. Once the brown sugar is added, blitz in the remaining butter.
5. Return the speed to high and mix until the dough reaches the intense stage of gluten development, about 2 minutes.
6. Remove the dough from the mixer and bulk-ferment until the dough has doubled in size, 20 to 30 minutes.
7. Divide the dough into 1 lb 8 oz/680 g pieces and preshape into rounds. Bench-rest the dough for about 15 minutes.
8. Using a rolling pin or a dough sheeter, roll the dough into a rectangle (12 × 24 inches) that is ¼ in/5 mm thick.

9. Smear the dough with poppy seed filling, making sure to leave a ½-in/1.25-cm border. Roll up the dough and press to close the seam.

10. Use a bench knife to cut the roll down the center lengthwise. Twist the 2 halves together, which will expose some of the filling.

11. Place the dough into greased and parchment-lined strap pans. Proof for 1½ to 2 hours, or as needed, depending on the environment. If desired, the loaves can be garnished with streusel before baking.

12. Load the bread into a 375°F/191°C convection oven and immediately drop the temperature to 350°F/177°C. Bake until golden brown, 30 to 40 minutes.

13. Immediately remove the finished bread from the pans to cool. To serve, brush the loaves with apricot glaze followed by the fondant. Or, the loaves may be garnished with confectioners' sugar after they have cooled.

CHEF'S NOTE

Though the Poppy Seed Filling is traditional, this bread can also be filled with the Chocolate Filling (page 218) or Cinnamon Filling (page 219).

Spread the filling evenly, to the very outside edges of the rolled dough.

After rolling the dough into a log, use a bench knife to cut the log in half.

Twist the two pieces together. You can choose to either expose or enclose the filling, depending on how the dough is twisted.

Babka is finished with apricot glaze, confectioners' sugar, and cake crumbs.

Poppy Seed Filling

YIELD: 3 LB 2-3/4 OZ/1.2 KG

INGREDIENT	US	METRIC
Milk	9 oz	250 g
Sugar	10½ oz	300 g
Butter	5½ oz	160 g
Poppy seeds, freshly ground	1 lb 2 oz	510 g
Cake crumbs	2 oz	60 g
Ground cinnamon	½ fl oz	15 ml
Lemon juice	½ oz	15 g
Salt	½ oz	15 g
Eggs, lightly beaten	5¼ oz	150 g

1. In a saucepan over medium heat, bring the milk, sugar, and butter to a boil.
2. Stir in the poppy seeds, cake crumbs, cinnamon, lemon juice, and salt. Remove from the heat and set aside to cool.
3. Once cooled, fold in the eggs. Cover and refrigerate until use.

Chocolate Filling

YIELD: 4 LB/2.3 KG

INGREDIENT	US	METRIC
Butter	1 lb 13 oz	825 g
Pastry flour	4 oz	110 g
Unsweetened cocoa powder	4 oz	110 g
Salt	Pinch	Pinch
Brown sugar	12 oz	790 g
Honey	3 oz	90 g
Eggs	12 oz	345 g
Vanilla extract	¼ oz	5 g

1. Prepare the filling the day before intended use. Melt the butter, and allow it to cool slightly.
2. Sift together the pastry flour, cocoa powder, and salt and set aside.
3. Combine the cooled butter with the brown sugar in a mixer. Mix with the paddle attachment until combined, scraping down the bowl often.
4. Add the sifted dry ingredients and mix until homogenous, scraping the bowl as needed. Add the honey and mix until combined, scraping the bowl as needed.
5. Combine the eggs and vanilla and add to the mixer. Mix until combined and smooth, scraping as needed.
6. Cover and refrigerate until use. Use while cold.

Cinnamon Filling

INGREDIENT	US	METRIC
Butter	1.82 lb	827 g
Brown sugar	1.74 lb	789 g
Pastry flour	4 oz	109 g
Cinnamon	4 oz	109 g
Honey	3 oz	89 g
Eggs	12 oz	345 g
Vanilla extract	¼ tsp	6 g

1. Melt the butter.
2. Combine the melted butter with the brown sugar in a 12 qt. mixer with a paddle attachment. Scrape down the sides during mixing.
3. Add the flour, cinnamon, and honey. Mix, scraping down the sides of the bowl.
4. Mix together the eggs and vanilla extract and blend into the mixture in the bowl.
5. Place the cinnamon filling in a covered container and refrigerate until needed.

Cinnamon Raisin Bread

DESIRED DOUGH TEMPERATURE: 76°F/24°C

INGREDIENT	US	METRIC	BAKERS' PERCENTAGE
Milk	5 lb 15 oz	2.7 kg	54.1%
Eggs	1 lb	455 g	9.1%
Butter	1 lb	455 g	9.1%
Malt syrup	1 oz	30 g	0.5%
Bread flour	11 lb	5 kg	100.0%
Yeast, instant dry	3 oz	85 g	1.7%
Sugar	1 lb	455 g	9.1%
Salt	4 oz	115 g	2.3%
Ground cinnamon	1 oz	30 g	0.5%
Raisins	2 lb 1 oz	910 g	18.2%
Total Dough Weight	22 lb 9 oz	10.2 kg	204.6%
Cinnamon Filling (page 219)	1 lb 14 oz	850 g	
Egg wash	As needed	As needed	

1. Combine the milk, eggs, butter, malt syrup, flour, yeast, sugar, and salt in a mixer and mix on low speed until homogenous, about 4 minutes.
2. Increase the speed to high and mix until the dough reaches the intense stage of gluten development, about 4 minutes.
3. Add the cinnamon and raisins, and mix on low speed to combine, about 1 minute.
4. Remove the dough from the mixer and bulk-ferment, covered, until doubled in size, about 45 minutes.
5. Divide the dough into 1 lb 2 oz/510 g pieces. Roll each piece into a 12 by 16-in/30 by 41-cm square.
6. Spread 1½ oz/43 g of filling onto the dough, leaving a ½-in/1-cm border on the edges.
7. Starting with the edge closest to you, roll the dough lengthwise. Use your palm to close the ends and seams.
8. Place the dough into well-oiled strap pans and brush with egg wash. Proof for 45 minutes to 1 hour, or as needed, depending on the environment.
9. Apply a second coat of egg wash. Bake the dough in a 365°F/185°C oven until golden brown, 25 to 30 minutes.
10. Remove breads from the pans after baking and place on a rack to cool.

Honey Whole Wheat Bread

DESIRED DOUGH TEMPERATURE: 76°F/24°C

INGREDIENT	US	METRIC	BAKERS' PERCENTAGE
Biga			
Water	1 lb	450 g	8.9%
Durum flour	7¾ oz	220 g	4.3%
Semolina flour	1 lb 4½ oz	580 g	11.6%
Yeast, instant dry	Pinch	0.3 g	—
Final Dough			
Water	6 lb 2½ oz	2.8 kg	55.0%
Biga	2 lb 10 oz	1.2 kg	24.9%
Olive oil	10¼ oz	290 g	5.8%
Malt syrup	½ oz	15 g	0.3%
Durum flour	5 lb 1 oz	2.3 kg	45.7%
Bread flour	4 lb 3 oz	1.9 kg	38.4%
Yeast, instant dry	¾ oz	20 g	0.4%
Salt	4¼ oz	120 g	2.3%
Total Dough Weight	19 lb	8.7 kg	172.8%

1. Prepare the biga (see page 125) 18 hours before mixing the final dough.
2. To prepare the final dough, combine the water, biga, oil, malt syrup, durum flour, and bread flour in a mixer, and mix on low speed until homogenous, about 3 minutes. Stop the mixer and rest the dough to autolyse for 15 to 20 minutes.
3. Add the yeast and mix on low speed to incorporate, about 1½ minutes. Add the salt and mix to combine, about 30 seconds. Increase the speed to high and mix until the dough reaches the improved stage of gluten development, about 2 minutes.
4. Remove the dough from the mixer and bulk-ferment, covered, for about 45 minutes. Fold and ferment for an additional 15 minutes.
5. Divide the dough into 1 lb 8 oz/680 g pieces and preshape into rounds. Bench-rest the dough for about 15 minutes.
6. Shape the dough into rounds and place into lightly floured round bannetons. Proof the dough for 30 to 40 minutes, or as needed, depending on the environment.
7. Score the dough and load into a 470°F/243°C oven and bake until golden brown about 20 minutes. Open the vent and bake for an additional 8 minutes.
8. Place the finished breads on a rack to cool.

CHEF'S NOTE

For Honey and Walnut Whole Wheat Bread, add roughly chopped walnuts, at 10% of the total dough weight.

Gugelhopf

DESIRED DOUGH TEMPERATURE: 78°F/26°C

INGREDIENT	US	METRIC	BAKERS' PERCENTAGE
Soaker			
Raisins	1 lb 12¼ oz	800 g	16.0%
Dried cherries	1 lb 5 oz	600 g	12.0%
Kirschwasser	10½ oz	300 g	6.0%
Final Dough			
Bread flour	11 lb	5 kg	100.0%
Yeast, fresh	8½ oz	240 g	4.8%
Milk, 80°F/27°C	5 lb 4½ oz	2.4 kg	48.0%
Eggs	2 lb 3 ¼ oz	1 kg	20.8%
Sugar	2 lb 3¼ oz	1 kg	20.8%
Butter	3 lb 12 oz	1.7 kg	34.4%
Salt	5 oz	140 g	2.8%
Soaker	7 lb 11 oz	3.5 kg	68.0%
Total Dough Weight	33 lb	15 kg	299.6%
Blanched whole almonds	2 lb	910 g	
Apricot glaze	As needed	As needed	

1. Prepare the soaker 18 hours before mixing the final dough.
2. To make the final dough, place the flour in a mixer and make a well. Dissolve the yeast in about half of the milk and add to the flour. By hand, mix in a small amount of the flour. Dust the mixture with flour and let rest until it cracks and forms craters, about 10 minutes.
3. Add the remaining milk, the eggs, sugar, butter, and salt. Mix on low speed until homogenous, about 3 minutes. Increase the speed to high and mix until the dough reaches the intense stage of gluten development, about 8 minutes.
4. Return the speed to low and add the soaker. Mix just until incorporated, about 1 minute.
5. Remove the dough from the mixer and bulk-ferment, covered, for about 2 hours. Punch to deflate twice during fermentation.
6. Divide the dough into 1 lb 14 oz/850 g pieces and preshape into rounds.

7. Lightly spray Gugelhopf molds and evenly space 10 almonds in the bottom of each mold, depending on the size.
8. Use a rolling pin to make a hole in the center of each round and place the rounds seam side up in the prepared molds. Proof the dough until doubled in volume, about 1 hour, or as needed, depending on the environment.
9. Bake in a 380°F/193°C oven until golden brown, about 35 minutes.
10. Cool completely before removing from molds. Brush with apricot glaze and fondant.

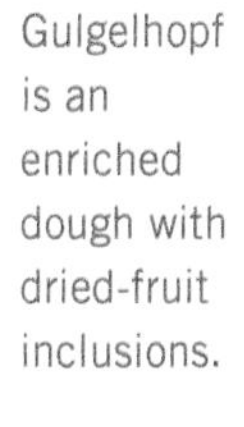

Gulgelhopf is an enriched dough with dried-fruit inclusions.

After shaping into rounds, use a rolling pin to make a hole in the center of each loaf.

Place the dough into oiled pans to proof.

Finished Gugelhopf is glazed with fondant and toasted almonds.

Oat and Fruit Bread

DESIRED DOUGH TEMPERATURE: 77°F/25°C

INGREDIENT	US	METRIC	BAKERS' PERCENTAGE
Soaker			
Water, hot	1 lb 2½ oz	525 g	9.5%
Rolled oats	1 lb 2½ oz	525 g	9.5%
Final Dough			
Water	7 lb 12 oz	3.5 kg	63.7%
Bread flour	9 lb 4 oz	4.2 kg	76.2%
Medium rye flour	1 lb 12 oz	790 g	14.3%
Rye Sourdough Starter (page 147)	1 lb ½ oz	465 g	8.4%
Soaker	2 lb 5 oz	1.1 kg	19.0%
Salt	4½ oz	130 g	2.4%
Yeast, instant dry	1¼ oz	35 g	0.6%
Honey	15 oz	420 g	7.6%
Malt syrup	1½ oz	35 g	0.6%
Pâte Fermentée (page 199)	2 lb 10 oz	1.2 kg	21.6%
Assorted diced fruit (apples, apricots, figs, dates)	5 lb 15 oz	2.7 kg	48.6%
Walnuts	1 lb 9 oz	715 g	13.0%
Total Dough Weight	33 lb 10¾ oz	15.29 kg	276.0%
Rolled oats	As needed	As needed	

1. Prepare the soaker at least 1 hour before mixing the final dough.

2. To mix the final dough, combine the water, bread flour, rye flour, rye sourdough starter, soaker, salt, yeast, honey, malt syrup, and pâte fermentée in a mixer, and mix on low speed until homogenous, about 4 minutes.

3. Increase the speed to high and mix until a substantial dough has formed, about 2 minutes.

4. Add the fruit mixture and mix on low speed until incorporated, about 2 minutes.

5. Transfer the dough to a tub and sprinkle the walnuts on top of the dough. Bulk-ferment, covered, for

45 minutes. Fold the dough and ferment for an additional 15 minutes.

6. Divide the dough into 3 lb 10 oz/1.7 kg presses and preshape into rounds. Bench-rest the dough for 10 to 15 minutes.
7. Use a dough divider to divide the dough into 18 pieces (if the divider will only cut 36 pieces, roll 2 pieces together to yield 18).
8. Shape the dough into 5-in/13-cm oblong logs and roll in the oats. Place on a parchment-lined sheet tray in 5 rows of 3 each. Proof the dough for 35 to 45 minutes, or as needed, depending on the environment.
9. Load the bread into a 470°F/243°C oven and immediately lower the temperature to 410°F/210°C. Bake until golden brown, about 16 minutes.
10. Place the finished breads on a rack to cool.

CHEF'S NOTE

Any fruit can be used in this bread. Dried fruits will absorb water from the dough, so soak them in water for 30 minutes before using.

Pain au Fromage

DESIRED DOUGH TEMPERATURE: 74°F/23°C

INGREDIENT	US	METRIC	BAKERS' PERCENTAGE
Water	6 lb 10 oz	3 kg	60.0%
Bread flour	11 lb	5 kg	100.0%
Liquid Levain (page 132)	5 lb 4½ oz	2.4 kg	48.5%
Yeast, instant dry	1 oz	30 g	0.5%
Sea salt	3 oz	90 g	1.8%
Extra-virgin olive oil	1 lb 2 oz	510 g	10.0%
Asiago, grated	2 lb	910 g	18.0%
Parmesan, grated	12 oz	340 g	6.8%
Total Dough Weight	27 lb	12.3 kg	245.6%

1. Combine the water, flour, liquid levain, and yeast in a mixer, and mix on low speed until homogenous, about 1½ minutes. Add the salt and mix to combine, about 30 seconds.
2. Increase the speed and mix on high until the dough reaches the improved stage of gluten development, about 2 minutes.
3. Return the speed to low and slowly add the oil. Add the cheeses and mix until incorporated, about 1 minute.
4. Remove the dough from the mixer and bulk-ferment, covered, until doubled in size, about 2 hours. Fold twice during fermentation.
5. Divide the dough into 1 lb 10 oz 750 g pieces and preshape into rounds. Bench-rest the dough for about 15 minutes.
6. Shape the dough into boules or bâtards and place seam side up in floured bannetons (see Note). Proof for about 2 hours or as needed, depending on the environment.
7. Score and load into a 470°F/243°C oven. Steam for 1 second before loading and 3 seconds after loading. Lower the temperature to 400°F/204°C and bake until golden brown, about 45 minutes.
8. Place finished breads on a rack to cool.

CHEF'S NOTE

This dough can also be shaped into rolls. Scale 3½ oz/ 100 g pieces and shape into rounds. Bake on parchment-lined sheet trays for 20 to 25 minutes.

Ham and Provolone Bread

DESIRED DOUGH TEMPERATURE: 78°F/26°C

INGREDIENT	US	METRIC	BAKERS' PERCENTAGE
Water	6 lb 8½ oz	3 kg	59.2%
Bread flour	11 lb	5 kg	100.0%
Yeast, instant dry	2 oz	50 g	1.0%
Salt	4 oz	110 g	2.2%
Olive oil	1 lb 2 oz	505 g	10.1%
Butter	5 oz	135 g	2.7%
Malt syrup	1½ oz	40 g	0.8%
Ham, medium diced	3 lb 8½ oz	1.6 kg	32.0%
Provolone, medium diced	3 lb 8½ oz	1.6 kg	32.0%
Total Dough Weight	26 lb 8 oz	12 kg	240.0%

1. Combine the water, flour, yeast, salt, oil, butter, and malt syrup in a mixer, and mix on low speed until homogenous, about 4 minutes. Increase the speed to high and mix until the dough reaches the improved stage of gluten development, about 2 minutes.
2. Add the ham and cheese and mix just to combine, about 1 minute.
3. Remove the dough from the mixer and bulk-ferment, covered, for 45 minutes. Fold and ferment for an additional 15 minutes.
4. Divide the dough into 1 lb 4 oz/570 g pieces and preshape into rounds. Bench-rest the dough for about 15 minutes.
5. Shape the dough into rounds and place into lightly floured bannetons. Proof for about 45 minutes or as needed, depending on the environment.
6. Line a loader with parchment paper before transferring the dough. Score the rounds with a crosshatch pattern.
7. Load the dough into a 460°F/238°C oven and steam during the first 2 minutes. Bake until lightly golden, about 20 minutes. Open the vent and bake until golden brown, 5 to 10 minutes.
8. Place the finished breads on a rack to cool.

Pesto Bread

DESIRED DOUGH TEMPERATURE: 78°F/26°C

INGREDIENT	US	METRIC	BAKERS' PERCENTAGE
Water	6 lb 6½ oz	2.9 kg	58.1%
Bread flour	11 lb	5 kg	100.0%
Yeast, instant dry	1¾ oz	50 g	1.0%
Salt	4 oz	115 g	2.2%
Malt syrup	1½ oz	40 g	0.8%
Butter, softened	5 oz	140 g	2.7%
Pesto	1 lb 9 oz	705 g	14.1%
Total Dough Weight	19 lb 11½ oz	3.9 kg	178.9%

1. Combine the water, flour, yeast, salt, malt syrup, butter, and pesto in a mixer and mix on low speed until homogenous, about 4 minutes. Increase the speed to high and mix until the dough reaches the intense stage of gluten development, about 2 minutes.
2. Remove the dough from the mixer and bulk-ferment, covered, for about 45 minutes. Fold and ferment for an additional 15 minutes.
3. Divide the dough into 1 lb 4 oz/570 g pieces and preshape into rounds. Bench-rest the dough for 10 to 15 minutes.
4. Shape the dough into 10-in/25-cm bâtards and place in lightly floured couches or oblong bannetons. Proof the dough for about 45 minutes or as needed, depending on the environment.
5. Score and load the dough into a 450°F/232°C oven and bake until golden brown, about 30 minutes.
6. Place the finished breads on a rack to cool.

Pesto

YIELD: 32 FL OZ/960 ML

INGREDIENT	US	METRIC
Toasted pine nuts	5 oz	142 g
Minced garlic	1 oz	28 g
Basil leaves	12 oz	340 g
Grated Parmesan	7 oz	198 g
Olive oil	6 fl oz	180 ml
Salt	2 tsp	6.5 g
Ground black pepper	2 tsp	4 g

1. Combine the pine nuts, garlic, basil, and Parmesan in a food processor fitted with the metal chopping blade. Process to blend.
2. Add the olive oil with the processor running and process until smooth. Adjust seasoning with salt and pepper. Cover and refrigerate until needed.

Pain de Brioche

DESIRED DOUGH TEMPERATURE: 77°F/25°C

INGREDIENT	US	METRIC	BAKERS' PERCENTAGE
Water	3 lb 8½ oz	1.6 kg	32.0%
Bread flour	11 lb	5 kg	100.0%
Milk powder	9 oz	250 g	5.0%
Eggs	2 lb 7 oz	1.1 kg	22.0%
Yeast, instant dry	7 oz	200 g	4.0%
Sugar	1 lb 5 oz	600 g	12.0%
Salt	3½ oz	100 g	2.0%
Butter, softened	5 lb 12 oz	2.6 kg	52.0%
Total Dough Weight	25 lb 4 oz	11.5 kg	229.0%
Egg wash	As needed	As needed	

1. Prepare the dough a day ahead. Combine the water, flour, milk powder, eggs, and yeast in a mixer, and mix on low speed until homogenous, about 2 minutes. Add the sugar and salt and mix until incorporated, about 1 minute. Increase the speed to high and mix until the dough reaches the intense stage of gluten development, about 3 minutes.
2. Return the mixer to low speed and slowly add the butter. Mix until incorporated, about 2 minutes.
3. If the ambient temperature is cold, leave the dough on plastic-wrapped sheet trays to proof for about 1 hour. Fold once, rewrap, and place in the refrigerator overnight. If the ambient temperature is warm, remove the dough from the mixer and proof for about 25 minutes. Fold the dough, rewrap the tray, and refrigerate overnight. If it is very hot (more than 85°F/29°C), refrigerate the dough immediately, without folding.
4. Divide the dough into 5¾ oz/165 g pieces and preshape into rounds. Bench-rest the dough for about 10 minutes.
5. Shape the dough into rounds and place 3 per pan in oiled loaf pans. Brush with egg wash. Proof the dough for about 1 hour or as needed, depending on the environment.
6. Apply a second coat of egg wash. Use scissors to cut 3 "X" scores in each loaf.
7. Bake in a 380°F/193°C oven until golden brown, about 35 minutes. Rotate the pans once during baking.
8. Cool in the pans for 5 minutes before inverting. Place the finished breads on a rack to cool.

Pain de Meteil

DESIRED DOUGH TEMPERATURE: 75°F/24°C

INGREDIENT	US	METRIC	BAKERS' PERCENTAGE
Water	6 lb 15 oz	3.2 kg	63.0%
All-purpose flour	5 lb 8 oz	2.5 kg	50.0%
Bread flour	5 lb 8 oz	2.5 kg	50.0%
Butter, softened	4¾ oz	135 g	2.7%
Powdered milk	3½ oz	100 g	2.0%
Yeast, instant dry	4¾ oz	135 g	2.7%
Sea salt	1 ¾ oz	50 g	1.0%
Sugar	5¼ oz	150 g	3.0%
Total Dough Weight	19 lb 1¼ oz	8.7 kg	174.4%

1. Combine the water, all-purpose flour, bread flour, butter, and powdered milk in a mixer, and mix on low speed until combined, about 1½ minutes. Sprinkle the yeast on top of the dough and rest in the mixer to autolyse for 15 to 20 minutes.
2. Resume mixing on low speed to incorporate the yeast. Add the salt and sugar and mix until combined, about 1 minute.
3. Increase the speed to high and mix until the dough reaches the improved stage of gluten development, about 3 minutes.
4. Remove the dough from the mixer and bulk-ferment, covered, until doubled in size, about 1 hour.
5. Divide the dough into 2 lb 3 oz/1 kg pieces and preshape into oblong logs. Bench-rest the dough for about 20 minutes.
6. Shape into 18-in/46-cm logs and place in lightly greased Pullman pans. Proof for about 1 hour (with the lids off) or as needed, depending on the environment.
7. Place the oiled lids on top of the Pullmans and load in a 400°F/204°C oven. Bake until golden brown, about 1 hour.
8. Remove from the pans and place finished breads on a rack to cool.

Pain de Mie

DESIRED DOUGH TEMPERATURE: 76°F/24°C

INGREDIENT	US	METRIC	BAKERS' PERCENTAGE
Water	7 lb 1 oz	3.2 kg	64.3%
Bread flour	11 lb	5 kg	100.0%
Yeast, instant dry	1½ oz	45 g	0.9%
Salt	4 oz	115 g	2.3%
Sugar	7¼ oz	205 g	4.1%
Butter	15½ oz	440 g	8.8%
Malt syrup	¾ oz	25 g	0.5%
Total Dough Weight	19 lb 14 oz	9 kg	180.9%

1. Combine the water, bread flour, yeast, salt, sugar, butter, and malt syrup in a mixer, and mix on low speed until homogenous, about 4 minutes. Increase the speed to high and mix until the dough reaches the intense stage of gluten development, about 4 minutes. The dough will be sticky, but will have good gluten development.
2. Remove the dough from the mixer and bulk-ferment, covered, until doubled in size, about 45 minutes.
3. The dough can be divided into small loaves or Pullman loaves. Divide into 1 lb 8 oz/680 g pieces for small loaves and 2 lb 8 oz/1.1 kg pieces for Pullman loaves.
4. Preshape into oblong logs and bench-rest the dough for 10 to 15 minutes.
5. For small loaves, shape the dough into 12-in/30-cm bâtards, and Pullmans into 18-in/46-cm bâtards and place into lightly oiled Pullman pans. Proof the dough for about 1½ hours (with the lids off for Pullmans) or as needed, depending on the environment.
6. Place the oiled lids on top of the Pullmans and load into a 400°F/204°C convection oven. Lower the temperature to 375°F/191°C and bake until golden brown, about 35 minutes. The internal temperature should read 205°F/96°C.
7. Remove the breads from the pans and place on a rack to cool.

Kaiser Rolls

DESIRED DOUGH TEMPERATURE: 77°F/25°C

INGREDIENT	US	METRIC	BAKERS' PERCENTAGE
Water	2 lb 10 oz	1.2 kg	52.0%
Bread flour	5 lb 1 oz	2.3 kg	100.0%
Malt syrup	1½ oz	45 g	2.0%
Oil	2 oz	55 g	2.4%
Yeast, instant dry	2½ oz	70 g	3.5%
Eggs	4 oz	115 g	5.5%
Sugar	1¾ oz	50 g	2.2%
Sea salt	2 oz	55 g	2.4%
Total Dough Weight	8 lb 8¾ oz	3.9 kg	170.0%
Vegetable oil	As needed	As needed	
Sesame seeds or poppy seeds (optional)	As needed	As needed	

1. Combine the water, flour, malt, oil, yeast, and eggs in a mixer. Mix on low speed until homogenous, about 45 seconds.
2. Stream in the salt and sugar and mix to combine, about 45 seconds.
3. Increase the speed and mix on high until the dough reaches the intense stage of gluten development, about 4 minutes.
4. Remove the dough from the mixer and bulk-ferment, covered, for 1 hour. Fold once halfway through fermentation.
5. Divide the dough into 4 oz/115 g pieces and pre-shape into rounds. Bench-rest the dough for 10 to 15 minutes.
6. Shape the dough into boules. Brush the top of each roll with oil, and stamp with a Kaiser roll stamp. Sprinkle with sesame seeds or poppy seeds, if using. Proof the rolls for about 1 hour or as needed, depending on the environment.
7. Load the dough into a 400°F/204°C convection oven and steam. Bake until golden brown, about 15 minutes.
8. Place the finished breads on a rack to cool.

CHEF'S NOTE

Though a stamp is more commonly used today, Kaiser rolls can also be hand-shaped. See page 235.

For high-volume production, a stamp can be used to create the signature star pattern on Kaiser Rolls.

Hand-shaped Kaiser Rolls are begun by folding half the dough to the center of the circle. Use your thumb as a guide to crimp the dough.

Continue to fold the dough to crimp the edges. Use the side of your hand to create a crease after each fold.

A stamped Kaiser Roll and a hand-shaped Kaiser Roll.

Hamburger Buns

DESIRED DOUGH TEMPERATURE: 76°F/24°C

INGREDIENT	US	METRIC	BAKERS' PERCENTAGE
Sponge			
Bread flour	5 lb 8 oz	2.5 kg	50.0%
Milk	4 lb 10 oz	2.1 kg	42.4%
Yeast, instant dry	1½ oz	40 g	0.8%
Final Dough			
Sponge	10 lb 5¾ oz	4.7 kg	93.2%
Bread flour	5 lb 8 oz	2.5 kg	50.0%
Milk	2 lb 7 oz	1.1 kg	22.3%
Butter	1 lb 4 oz	570 g	11.4%
Honey	13½ oz	380 g	7.6%
Malt syrup	1 oz	30 g	0.5%
Salt	4 oz	115 g	2.3%
Total Dough Weight	20 lb 11¼ oz	9.4 kg	187.3%
Egg wash	As needed	As needed	
Sesame seeds	As needed	As needed	

1. Mix the sponge the same day as the final dough. Combine the flour, milk, and yeast in a mixer, and mix on low speed until homogenous, about 2 minutes. Increase the speed to high and mix for an additional 1 minute. Ferment for 30 minutes.
2. To prepare the final dough, combine the sponge, flour, milk, butter, honey, malt syrup, and salt in a mixer. Mix on low speed until homogenous, about 4 minutes. Increase the speed and mix until the intense stage of gluten development is reached, about 4 minutes.
3. Remove the dough from the mixer and bulk-ferment, covered, until doubled in size, about 45 minutes.
4. Divide the dough into 5 lb/2.3 kg presses and pre-shape into rounds. Bench-rest the dough for 10 to 15 minutes.

5. Use a dough divider to divide the presses into 18 pieces (4½ oz/130 g per piece). If the divider only cuts 36 pieces, combine 2 pieces and shape into tight rounds.

6. Place the rolls on a parchment-lined sheet tray in 4 rows of 2 each, and brush with egg wash. Proof the dough for 20 minutes or as needed, depending on the environment.

7. Gently flatten the rolls and apply a second coat of egg wash. Sprinkle the top of each bun with sesame seeds.

8. Load the dough into a 375°F/191°C oven and bake until golden brown, about 15 minutes.

9. Place the finished buns on a rack to cool.

Honey Wheat Buns

DESIRED DOUGH TEMPERATURE: 76°F/24°C

INGREDIENT	US	METRIC	BAKERS' PERCENTAGE
Milk	9 lb 8 oz	4.3 kg	77.0%
Bread flour	11 lb	5 kg	89.7%
Yeast, instant dry	1¼ oz	35 g	0.6%
Wheat bran	1 lb 4½ oz	580 g	10.3%
Salt	4½ oz	125 g	2.2%
Vegetable oil	10¼ oz	290 g	5.2%
Honey	1 lb 1 oz	485 g	8.6%
Total Dough Weight	23 lb 13½ oz	10.8 kg	193.6%
Egg wash	As needed	As needed	

1. Combine the milk, flour, yeast, wheat bran, salt, oil, and honey in a mixer, and mix on low speed until homogenous, about 4 minutes. Increase the speed to high and mix until the dough reaches the intense stage of gluten development, about 6 minutes.
2. Remove the dough from the mixer and bulk-ferment for about 45 minutes.
3. Divide the dough into 6 lb/2.7 kg pieces and pre-shape into round presses. Bench-rest the dough for about 15 minutes.
4. Use a dough divider to divide the presses into 18 pieces. Place the rounds onto a parchment-lined sheet tray. Brush the buns with the egg wash. Proof the dough for about 45 minutes or as needed, depending on the environment.
5. Lightly flatten the buns with the palm of your hand and apply a second coat of egg wash.
6. Load the buns into a 370°F/188°C oven and bake until golden brown about 15 minutes.
7. Cool the buns completely before packing or serving.

Soft Rolls

DESIRED DOUGH TEMPERATURE: 78°F/26°C

INGREDIENT	US	METRIC	BAKERS' PERCENTAGE
Milk	5 lb 8 oz	2.5 kg	50.2%
Eggs	1 lb	455 g	9.1%
Malt syrup	¾ oz	20 g	0.4%
Bread flour	11 lb	5 kg	100.0%
Yeast, instant dry	2½ oz	75 g	1.5%
Salt	3¾ oz	110 g	2.2%
Butter, softened	1 lb	455 g	9.1%
Sugar	1 lb	455 g	9.1%
Total Dough Weight	20 lb	9.1 kg	181.6%
Egg wash	As needed	As needed	

1. Combine the milk, eggs, malt syrup, flour, yeast, salt, butter, and sugar in a mixer, and mix on low speed until homogenous, about 4 minutes. Increase the speed to high and mix until the dough reaches the intense stage of gluten development, about 4 minutes.
2. Remove the dough from the mixer and bulk-ferment, covered, until doubled in size, about 45 minutes.
3. Divide the dough into 1¾ oz/50 g pieces or into a 4 lb/1.8 kg press to divide on a dough divider, yielding 36 rolls.
4. Preshape the rolls into rounds and bench-rest the dough for 10 to 15 minutes.
5. Shape the rolls into rounds and place on a parchment-lined sheet tray in 5 rows of 3 each (see Note). Brush the rolls with the egg wash. Proof the rolls for 35 to 45 minutes or as needed, depending on the environment.
6. Apply a second coat of egg wash.
7. Load into a 390°F/199°C convection oven and bake until golden brown, about 14 minutes.
8. Cool the rolls on the sheet trays before transferring.

CHEF'S NOTE

This dough can be used to make knot rolls. Pre-shape the pieces into oblong logs and bench-rest for 10 to 15 minutes. Tie the rolls into knots and place on a parchment-lined sheet tray. Brush with egg wash, proof the knot rolls for 35 to 45 minutes, and proceed with Step 6 of the recipe.

Conchas

DESIRED DOUGH TEMPERATURE: 78°F/26°C

INGREDIENT	US	METRIC	BAKERS' PERCENTAGE
Sponge			
Bread flour	5 lb 8 oz	2.5 kg	50.0%
Milk	5 lb 1 oz	2.3 kg	46.6%
Yeast, instant dry	3 oz	80 g	1.6%
Final Dough			
Sponge	10 lb 12¾ oz	4.9 kg	98.2%
Bread flour	5 lb 8 oz	2.5 kg	50.0%
Malt syrup	1 oz	30 g	0.5%
Eggs	1 lb 14 oz	850 g	17.0%
Salt	3½ oz	100 g	2.0%
Butter, softened	2 lb 3¼ oz	1 kg	20.0%
Sugar	3 lb 5 oz	1.5 kg	30.0%
Total Dough Weight	24 lb	10.9 kg	217.7%
Vanilla Concha Topping (page 243) or Chocolate Concha Topping (page 242)	As needed	As needed	

1. Prepare the sponge and the final dough a day ahead of baking. Combine the flour, milk, and yeast in a mixer and mix on low speed until homogenous. Increase the speed to high and mix until the dough comes together, about 1 minute. Ferment for 30 minutes.

2. To prepare the final dough, combine the sponge, flour, malt syrup, eggs, salt, and about one third of the butter in a mixer. Mix on low speed until homogenous, about 5 minutes. Increase the speed to high and mix until the dough has reached the intense stage of gluten development, about 2 minutes.

3. Add another third of the butter while mixing on high speed, scraping often. Stream in half the sugar and then add the remaining butter. Stream in the remaining sugar and mix the dough on high until it reaches the intense stage of gluten development.

4. Bulk-ferment, covered, until the dough has doubled in size, about 1 hour.

5. Divide the dough into 2 oz/55 g pieces (or if using a dough divider, a 4 lb 8 oz/2 kg press, for 36 pieces). Bench-rest the dough for 10 to 15 minutes.

6. Shape the dough into tight rounds and place on a parchment-lined sheet tray in 5 rows of 3 each. Cover with plastic wrap and place the dough in the refrigerator to retard overnight.
7. Remove the dough from the refrigerator and top with the vanilla or chocolate topping. Proof for 45 minutes to 1 hour or as needed, depending on the environment.
8. Load the dough into a 375°F/191°C convection oven and immediately lower the temperature to 350°F/177°C. Bake until the conchas are golden brown, about 13 minutes.
9. Place the finished conchas on a rack to cool.
10. Spray the tops of the conchas with water and gently place the topping rounds onto the conchas, pressing slightly to adhere them. Use a concha cutter to mark the topping.

Chocolate Concha Topping

YIELD: 2 LB 3¾ OZ/1.0 KG

INGREDIENT	US	METRIC	BAKERS' PERCENTAGE
Cake flour	13 oz	370 g	100.0%
Confectioners' sugar	10½ oz	300 g	80.4%
Unsweetened cocoa powder	1 oz	30 g	8.6%
Butter, softened	5 oz	150 g	40.3%
Shortening	5½ oz	160 g	42.7%

1. Sift together the flour, confectioners' sugar, and cocoa powder.
2. In a mixer fitted with the paddle attachment, cream the butter and shortening until light and fluffy, about 4 minutes.
3. Add the dry ingredients and mix until homogenous, about 1 minute.
4. Remove from the mixer and wrap the dough in plastic. Refrigerate overnight.
5. Roll the dough until ¼ in/6 mm thick and cut with a 4-in/10-cm round cutter. Keep rounds refrigerated until topping is needed.
6. Excess topping can be wrapped and refrigerated until the next use.

CHEF'S NOTE

This topping can be prepared the same day as the final dough and used the following day to finish the conchas.

Vanilla Concha Topping

YIELD: 2 LB 4½ OZ/1.0 KG

INGREDIENT	US	METRIC	BAKER'S PERCENTAGE
Cake flour	14 oz	400 g	100.0%
Confectioners' sugar	10½ oz	300 g	74.9%
Vanilla extract	1 oz	30 g	8.0%
Butter, softened	2¾ oz	75 g	19.3%
Shortening	8 oz	225 g	56.9%

1. Sift together the flour and confectioners' sugar.
2. In a mixer fitted with the paddle attachment, cream the vanilla, butter, and shortening until light and fluffy, about 4 minutes.
3. Add the dry ingredients and mix until homogenous, about 1 minute.
4. Remove from the mixer and wrap the dough in plastic. Refrigerate overnight, or until needed.
5. Roll the dough until ¼ in/6 mm thick and cut with a 4-in/10-cm round cutter. Keep rounds refrigerated until topping is needed.
6. Excess topping can be wrapped and refrigerated until the next use.

CHEF'S NOTE

This topping can be prepared the same day as the final dough and used the following day to finish the conchas.

The flavoring in this dough can be changed to reflect seasonal flavors and ingredients.

Hot Cross Buns

DESIRED DOUGH TEMPERATURE: 76°F/24°C

INGREDIENT	US	METRIC	BAKERS' PERCENTAGE
Sponge			
Milk	5 lb 8 oz	2.5 kg	50.0%
Malt syrup	1 oz	30 g	0.5%
Bread flour	4 lb 6½ oz	2 kg	39.9%
Yeast, instant dry	4½ oz	125 g	2.5%
Final Dough			
Eggs	1 lb 11 oz	765 g	15.3%
Lemon zest, grated	2½ oz	75 g	1.5%
Butter	1 lb 11 oz	765 g	15.3%
Honey	5½ oz	160 g	3.2%
Sponge	10 lb 2½ oz	4.6 kg	92.4%
Bread flour	6 lb 13 oz	3.1 kg	60.1%
Sugar	1 lb 11 oz	765 g	15.3%
Salt	4½ oz	125 g	2.5%
Ground cinnamon	½ tsp	1 g	—
Grated nutmeg	2½ tsp	5 g	0.1%
Ground allspice	1 tsp	2 g	—
Dried black currants	3 lb 5 oz	1.5 kg	30.0%
Candied lemon peel, diced small	1 lb 5 oz	605 g	12.1%
Total Dough Weight	27 lb 5 oz	12.4 kg	247.8%

Egg wash	As needed	As needed
Hot Cross Bun Topping (page 246)	As needed	As needed
Apricot glaze	As needed	As needed
Pistachios, toasted and chopped	1 lb 4 oz	570 g

1. Prepare the sponge on the same day as the final dough. Combine the milk, malt syrup, flour, and yeast in a mixer, and mix on low speed until homogenous, about 2 minutes. Increase the speed to high and mix for about 1 minute. Ferment for 30 minutes.
2. To prepare the final dough, combine the eggs, zest, butter, honey, sponge, flour, sugar, salt, cinnamon, nutmeg, and allspice in a mixer. Mix on low speed until homogenous, about 4 minutes. Increase the speed to high and mix until the dough reaches the intense stage of gluten development, about 4 minutes.
3. Add the currants and lemon peel, and mix on low speed until incorporated, about 1 minute.
4. Remove the dough from the mixer and bulk-ferment, covered, for about 45 minutes.
5. Divide the dough into 4 lb 8 oz/2 kg presses and preshape into tight rounds. Bench-rest the dough for 10 to 15 minutes.
6. Divide the dough into 36 pieces on a dough divider (each roll should be 2 oz/57 g).
7. Place the rolls on a parchment-lined sheet tray in 8 rows of 6 each. Brush with egg wash. Proof the dough for 20 to 25 minutes or as needed, depending on the environment.
8. Apply a second coat of egg wash and allow the rolls to dry slightly, about 5 minutes. Use a #2 round tip to pipe the topping in the shape of a cross. Proof for an additional 20 to 25 minutes.
9. Load into a 375°F/191°C oven and bake until golden brown, about 18 minutes.
10. Remove the buns from the oven, immediately brush with apricot glaze, and garnish with the pistachios. Cool completely.

Hot Cross Bun Topping

YIELD: 1 LB 9½ OZ/725 G

INGREDIENT	US	METRIC
Butter, melted	9½ oz	270 g
Sugar	8 oz	225 g
Eggs, room temperature	1½ oz	45 g
Milk	6 oz	170 g
Vanilla extract	½ oz	15 g
Lemon zest, grated	½ oz	15 g
Pastry flour	1 lb	455 g

1. Prepare the topping the same day as the final dough.
2. In a small mixer fitted with a paddle attachment, combine the butter, sugar, eggs, milk, vanilla, and zest. Mix until homogenous, scraping often, about 2 minutes.
3. Add the flour, mixing to combine, about 1 minute.

Chocolate and Oat Hot Cross Buns

DESIRED DOUGH TEMPERATURE: 76°F/24°C

INGREDIENT	US	METRIC	BAKERS' PERCENTAGE
Sponge			
Milk	5 lb 8 oz	2.5 kg	50.0%
Malt syrup	1 oz	30 g	0.5%
Bread flour	4 lb 6½ oz	2 kg	39.9%
Yeast, instant dry	4¼ oz	125 g	2.5%
Soaker			
Milk	2 lb ¾ oz	925 g	18.5%
Rolled oats	1 lb 11 oz	765 g	15.3%
Final Dough			
Eggs	1 lb 11 oz	765 g	15.3%
Lemon zest purée	2½ oz	75 g	1.5%
Butter	1 lb 11 oz	765 g	15.3%
Honey	5½ oz	160 g	3.2%
Sponge	10 lb 5¾ oz	4.7 kg	92.9%
Bread flour	6 lb 10 oz	3 kg	60.1%
Salt	4¼ oz	125 g	2.5%
Sugar	1 lb 11 oz	765 g	15.3%
Ground cinnamon	1 tsp	2 g	—
Grated nutmeg	2½ tsp	5 g	0.1%
Ground allspice	1 tsp	2 g	—
Semisweet chocolate (60%)	2 lb 10 oz	1.2 kg	24.6%
Rolled oats, toasted	1 lb 10 oz	740 g	14.8%
Total Dough Weight	27 lb 1¼ oz	12.3 kg	245.6%

Egg wash	As needed	As needed
Chocolate Oat Hot Cross Bun Topping (page 248)	As needed	As needed
Apricot glaze, heated	As needed	As needed
Pistachios, toasted and roughly chopped	As needed	As needed

CHEF'S NOTE

Hot Cross Buns can also be made into mini buns. Divide a 3 lb/1.4 kg press and place 10 by 7 on a parchment-lined standard sheet tray.

1. Mix the sponge the same day as the final dough. Combine the sponge ingredients in a mixer, and mix on low speed for 2 minutes, followed by 1 minute on high speed. Allow to ferment for 30 minutes.
2. While the sponge is fermenting, prepare the soaker by combining the milk and oats.
3. To mix the final dough, combine the soaker, eggs, lemon zest, butter, honey, sponge, flour, salt, sugar, cinnamon, nutmeg, and allspice in a mixer. Mix on low speed for 4 minutes, followed by high speed for 4 minutes or until intense gluten stage is reached.
4. Add the chocolate and toasted oats, and mix on low speed for 1 minute or until all ingredients are incorporated.
5. Remove the dough from the mixer and bulk-ferment for 45 minutes.
6. Divide the dough into a 4 lb 8 oz/2.1 kg press for a Dutchess or Erika divider, or divide individual rolls into 2 oz/60 g pieces (see Note). Preshape the dough into a tight round. Allow the dough to bench-rest for 10 to 15 minutes.
7. Divide on the Dutchess or Erika divider, if using, and place in 6 rows of 8 rolls each on a parchment-lined sheet tray. Brush with egg wash and place into the proofer for 15 minutes.
8. Brush the rolls again with egg wash and use a #2 round tip to pipe on the topping. Allow to proof for an additional 15 to 20 minutes.
9. Bake in a convection oven at 375°F/191°C for 18 minutes.
10. Remove the buns from the oven and allow to cool slightly. Brush with hot apricot glaze and top with pistachios. Cool completely.

Chocolate Oat Hot Cross Bun Topping

YIELD: 2 LB 12½ OZ/1.3 KG

INGREDIENT	US	METRIC	BAKERS' PERCENTAGE
Butter, melted	9½ oz	270 g	59.9%
Brown sugar	9½ oz	270 g	59.9%
Eggs	1½ oz	45 g	9.9%
Milk	7¼ oz	205 g	44.9%
Vanilla extract	½ oz	15 g	3.1%
Ground cinnamon	2½ tsp	5 g	1.1%
Pastry flour	1 lb	455 g	100.0%

1. Prepare the topping the same day as the Chocolate Oat Hot Cross Buns.
2. Combine the butter, sugar, eggs, milk, vanilla, and cinnamon in a mixer. Mix with a paddle attachment until homogenous, about 2 minutes. Scrape the bowl as needed.
3. Add the flour and mix until combined, about 2 minutes, scraping as needed.

Panettone

DESIRED DOUGH TEMPERATURE: 76°F/24°C

INGREDIENT	US	METRIC	BAKERS' PERCENTAGE
Sponge			
Bread flour	3 lb 12 oz	1.7 kg	33.1%
Milk	2 lb 10 oz	1.2 kg	24.0%
Yeast, instant dry	3½ oz	100 g	2.0%
Final Dough			
Bread flour	7 lb 4½ oz	3.3 kg	66.9%
Milk	2 lb 7 oz	1.1 kg	22.9%
Eggs	2 lb 14 oz	1.3 kg	25.3%
Orange zest, grated	1¼ oz	35 g	0.7%
Lemon zest, grated	1¼ oz	35 g	0.7%
Glucose syrup	7 oz	195 g	3.9%
Malt syrup	¾ oz	20 g	0.4%
Sponge	6 lb 10 oz	3 kg	59.0%
Yeast, instant dry	1¼ oz	35 g	0.7%
Butter, softened	2 lb 3¼ oz	1 kg	20.5%
Sugar	1 lb 10 oz	740 g	14.8%
Salt	5 oz	140 g	2.8%
Dark raisins	1 lb 6 oz	630 g	12.6%
Golden raisins	1 lb 6 oz	630 g	12.6%
Candied lemon peel, finely diced	1 lb 6 oz	630 g	12.6%
Candied orange peel, finely diced	1 lb 6 oz	630 g	12.6%
Total Dough Weight	28 lb 3¼ oz	12.8 kg	256.4%

Egg wash	As needed	As needed	
Butter, divided into 30 equal pieces	1 lb	455 g	

1. Mix the sponge the same day as the final dough. Combine the flour, milk, and yeast in a mixer, and mix on low speed until homogenous, about 2 minutes. Increase the speed and mix on high for an additional 1 minute. Transfer to an oiled bowl and ferment for 30 minutes.

2. To prepare the final dough, combine the flour, milk, eggs, orange zest, lemon zest, glucose syrup, malt syrup, sponge, yeast, and one third of the butter in a mixer. Mix on low speed until homogenous, about 4 minutes. Increase the speed and mix on high until the dough reaches the intense stage of gluten development, about 6 minutes.

3. Reduce the speed to low and add another third of the butter, followed by half the sugar. Add the remaining butter and remaining sugar. Mix the dough on high speed until it returns to the intense stage of gluten development, about 2 minutes.

4. Add the dark and golden raisins, candied orange peel, and candied lemon peel and mix on low speed until incorporated, about 1 minute.

5. Remove the dough from the mixer and bulk-ferment, covered, until doubled in size, about 1 hour.

6. Divide the dough into 1 lb 4 oz/565 g pieces. Shape the dough into tight rounds and place in oiled round pan molds. Brush the tops with the egg wash. Proof for 1 to 1½ hours or as needed, depending on the environment.

7. Apply a second coat of egg wash and score the top of the rounds with an "X." Place a piece of butter on the top of the pieces.

8. Bake in a 375°F/191°C convection oven until golden brown, 15 to 17 minutes.

9. Place the finished panettones on racks to cool.

Chocolate and Orange Panettone

DESIRED DOUGH TEMPERATURE: 76°F/24°C

INGREDIENT	US	METRIC	BAKERS' PERCENTAGE
Sponge			
Bread flour	3 lb 12 oz	1.7 kg	34.0%
Milk	2 lb 10 oz	1.2 kg	24.7%
Yeast, instant dry	3½ oz	100 g	2.0%
Ganache			
Water, boiling	13 oz	365 g	7.3%
Bittersweet chocolate, chopped	1 lb 12½ oz	805 g	16.1%
Final Dough			
Bread flour	7 lb 4½ oz	3.3 kg	66.0%
Milk	2 lb 2½ oz	980 g	19.6%
Eggs	2 lb 14 oz	1.3 kg	25.9%
Orange zest, grated	3½ oz	100 g	2.0%
Glucose syrup	13 oz	365 g	7.3%
Malt syrup	1 oz	30 g	0.5%
Yeast, instant dry	1¼ oz	35 g	0.7%
Salt	5 oz	145 g	2.9%
Butter	2 lb 10 oz	1.2 kg	24.4%
Sugar	1 lb 10¾ oz	760 g	15.2%
Ganache	2 lb 9½ oz	1.2 kg	24.0%
Bittersweet chocolate, chopped	2 lb 14 oz	1.3 kg	25.9%
Candied orange peel, diced small	2 lb 14 oz	1.3 kg	25.9%
Sponge	6 lb 10 oz	3 kg	60.6%
Total Dough Weight	32 lb 15.8 oz	15 kg	299.3%

Egg wash	As needed	As needed	
Butter, divided into 30 equal pieces	1 lb	455 g	

1. Prepare the sponge the same day as the final dough. Combine the flour, milk, and yeast in a mixer, and mix on low speed until homogenous, about 2 minutes. Increase the speed to high and mix an additional 1 minute. Ferment for 30 minutes.
2. Prepare the ganache by pouring the boiling water over the chopped chocolate. Stir until the chocolate is melted. Cool before use.
3. To prepare the final dough, combine the flour, milk, eggs, zest, glucose syrup, malt syrup, yeast, salt, and about one third of the butter in a mixer. Mix on low speed until homogenous, about 4 minutes. Increase the speed to high and mix until the dough reaches the intense stage of gluten development, about 6 minutes.
4. Return the speed to low and add the ganache in 2 additions. Scrape as needed. Mix until fully combined, about 1 minute.
5. Add in another third of the butter, followed by half the sugar. Mix on high speed until incorporated, about 2 minutes. Add the remaining butter and sugar, and mix until the dough returns to the intense stage of gluten development, about 4 minutes.
6. Add the chocolate and candied orange peel, and mix on low speed just to incorporate, scraping as needed.
7. Remove the dough from the mixer and bulk-ferment, covered, for 1 hour.
8. Divide the dough into 1 lb 3 oz/565 g pieces. Shape the dough into tight rounds and place in oiled round pan molds. Brush the tops with egg wash. Proof for 1 to 1½ hours or as needed, depending on the environment.
9. Apply a second coat of egg wash and score the top of the panettones with an "X." Place a piece of butter on the top of each.
10. Bake in a 375°F/191°C convection oven until golden brown, 15 to 17 minutes.
11. Place the finished panetonnes on racks to cool.

VARIATION

For a Chocolate Almond Panettone, omit the orange zest and candied orange peel. Add 2 lb 14 oz/1.3 kg slivered, toasted almonds when adding the chocolate in Step 6. Proceed with the recipe as written.

For a Chocolate Cherry Panettone, substitute 2 lb 14 oz/1.3 kg dried cherries for the candied orange peel.

Danish

DESIRED DOUGH TEMPERATURE: 77°F/25°C

INGREDIENT	US	METRIC	BAKERS' PERCENTAGE
Dough			
Milk	4 lb 7 oz	2 kg	43.3%
Eggs	2 lb 1 oz	930 g	19.9%
Malt syrup	¾ oz	20 g	0.4%
Bread flour	10 lb 5¾ oz	4.7 kg	100.0%
Yeast, instant dry	1½ oz	40 g	0.9%
Sugar	1 lb 2½ oz	525 g	11.3%
Salt	3½ oz	100 g	2.2%
Butter, pliable	14 oz	400 g	8.7%
Roll-in Butter			
Bread flour	12 oz	345 g	7.4%
Butter, pliable	7 lb 11 oz	3.5 kg	75.3%
Total Dough Weight	26 lb 11 oz	12.6 kg	269.4%
Fillings of choice (recipes follow)	As needed	As needed	

1. Prepare the dough a day ahead. Combine the milk, eggs, malt syrup, flour, yeast, sugar, salt, and butter in a mixer. Mix on low speed until homogenous, about 4 minutes, scraping often. Increase the speed to high and mix until the dough reaches the intense gluten stage of gluten development, about 4 minutes.
2. Remove the dough from the mixer and bulk-ferment, covered, until doubled in size, about 1 hour.
3. Roll the dough to a 13 by 18-in/33 by 46-cm rectangle. Cover and place in the refrigerator.
4. Prepare the roll-in butter by kneading the flour into the butter until it is well combined and pliable.
5. Fold a piece of parchment paper so that it is about 6 by 9 in/15 by 23 cm. Enclose the butter mixture in the parchment and use a rolling pin to roll the butter to the dimensions of the parchment package. Refrigerate the roll-in butter with the dough.
6. Retard the covered dough in the refrigerator overnight.
7. Remove the roll-in butter from the refrigerator and allow it to become more pliable.

8. Enclose the butter in the dough (to laminate, see page 207). Rest for 30 minutes in the refrigerator.
9. Roll the dough to a rectangle 12 by 28 in/31 by 71 cm and give the dough a three-fold: fold one end of the dough just past the center, and then fold the other half over the first, like a letter (see page 208). There should be 3 layers visible. Place the dough in the refrigerator to rest for 30 minutes.
10. Roll the dough again to a rectangle 12 by 28 in/31 by 71 cm and give the dough a four-fold: fold one end of the dough a quarter of the way. Bring the other end of the dough three-fourths of the way to meet the first edge to create a seam. Now fold the entire piece of dough in half, like a book (see page 208). There should be 4 layers visible. Place the dough in the refrigerator to rest for 30 minutes.
11. Roll the dough yet again to a rectangle 12 by 28 in/31 by 71 cm and give the dough a final three-fold. Wrap the dough in plastic wrap and freeze for 2 hours. Remove from the freezer and refrigerate overnight.
12. Allow the dough to come to working temperature. Roll the dough to a ¼-in/6-mm thickness, then fill and shape as desired.

Roll the dough into a rectangle 5 in/13 cm wide. Cut 1-in/2-cm strips of dough, and cut the strip into thirds, leaving the top one third intact. Roll the two outside pieces away from the center, followed by the center strip.

Roll the dough into a square, and cut 4-in/10-cm squares. Use a 2½-in/6-cm fluted circle cutter and cut as many circles as you have squares. Fill the squares, brush the border with egg wash, and fold the corners in to the meet at the center. Brush the tops with egg wash and place the circles in the center.

Roll the dough into a 12-in/30½-cm square. Spread the filling to the edges of the dough, and roll each side in to meet in the center. Cut the rolled dough into 1½-in/4-cm slices.

Assorted finished Danish.

Roll 4 lb/1.8 kg to a rectangle that measures 15 by 7 in/38 by 18 cm and 1/8 in/4 mm thick. Spread the desired filling on the dough, leaving a ½-in/1-¼-cm border without filling. Brush the border with egg wash and roll up lengthwise, like a Swiss roll, with the egg-washed border on the outside of the roll.

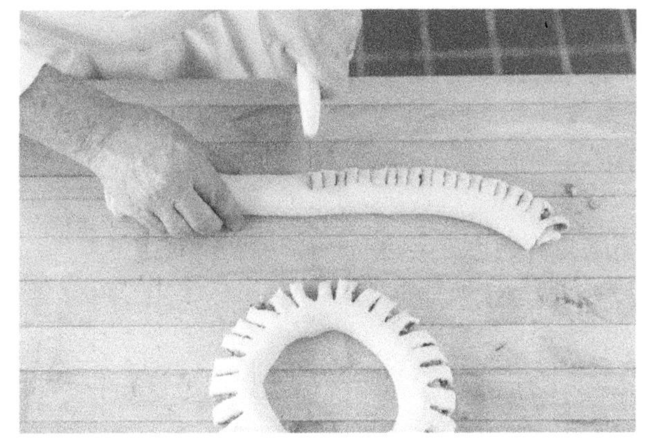

Using a bench knife, make cuts that go halfway through the diameter of the roll and are ½ in/1-¼ cm apart. Connect the two ends by tucking one end of the roll into the other to create a ring without a noticeable seam. Proof the dough as needed, depending on the environment. Brush the dough with egg wash and place the ring in a 375°F/191°C oven; immediately drop the temperature to 350°F/177°C. Bake until the Danish is golden brown and cooked through. Brush the dough with apricot glaze when it is removed from the oven. Allow the Danish to cool and then glaze the dough with 100°F/38°C poured fondant. Sprinkle chopped pistachios around the inner edge of the ring.

Finished Danish Ring

Almond Danish Filling

YIELD: 2 LB 5½ OZ/1.1 KG

INGREDIENT	US	METRIC
Almond paste	1 lb 1½ oz	500 g
Sugar	7 oz	200 g
Butter	5¼ oz	150 g
Orange zest, grated	2 oz	60 g
Lemon zest, grated	1½ oz	40 g
Eggs	3½ oz	100 g
Salt	½ oz	15 g
Lemon juice	½ oz	15 g

1. In the bowl of a mixer, combine the almond paste, sugar, butter, orange zest, lemon zest, eggs, salt, and lemon juice. Mix with the paddle attachment until homogenous, about 4 minutes. Scrape as needed.
2. Store the filling in the refrigerator until ready to use.

Apple Danish Filling

YIELD: 10 LB 10½ OZ/4.8 KG

INGREDIENT	US	METRIC
Apple cider	2 lb 12 oz	1.2 kg
Sugar	13 oz	365 g
Ground cinnamon	½ oz	10 g
Grated nutmeg	2½ tsp	5 g
Lemon juice	½ fl oz	15 ml
Clear gel starch	4 oz	120 g
Apples, medium diced	6 lb 12 oz	3.1 kg

1. In a saucepan, bring about 90 percent of the apple cider, the sugar, cinnamon, nutmeg, and lemon juice to a boil.
2. Stir together the remaining apple cider and the starch to form a slurry.
3. Stream the starch mixture into the saucepan, whisking constantly. Return to a boil and cook until thickened, about 1 minute.
4. Add the apples and cook until the apples begin to soften, about 1 minute.
5. Cool completely before using.

Blueberry Danish Filling

YIELD: 2 LB/910 G

INGREDIENT	US	METRIC
Blueberries, frozen, defrosted	1 lb 2 oz	510 g
Blueberry juice (see Note)	10½ oz	300 g
Sugar	2½ oz	75 g
Salt	Pinch	1 g
Clear gel starch	1 oz	30 g
Lemon juice	½ fl oz	15 ml

1. Place the blueberries in a strainer over a bowl. Allow to drain for 1 hour. Reserve the juice.
2. Bring 9 oz/255 g of the blueberry juice (see Note), the sugar, and salt to a boil.
3. Meanwhile, combine the remaining blueberry juice with the clear gel starch and mix until smooth.
4. Whisk the starch mixture into the boiling mixture, stirring constantly. Return to a boil and cook for 1 minute.
5. Add the blueberries and lemon juice and cook, stirring constantly, about 1 minute.
6. Remove from the heat and cool completely.
7. Cover and refrigerate until needed.

CHEF'S NOTE

If you do not have enough blueberry juice, make up difference with water or another juice, such as apple or cranberry.

Cherry Danish Filling

YIELD: 5 LB/2.4 KG

INGREDIENT	US	METRIC
Cherry juice	1 lb 8 oz	680 g
Sugar	7 oz	200 g
Salt	¾ oz	2 g
Clear gel starch	3 oz	75 g
Frozen cherries, drained	3 lb	1.4 kg

1. In a rondeau over moderate heat, bring 1 lb/450 g of the cherry juice, the sugar, and salt to a boil.
2. In a separate bowl, combine the remaining cherry juice with the clear gel starch and mix until smooth.
3. Whisk the starch mixture into the boiling cherry juice mixture, stirring constantly. Return to a boil and cook for about 1 minute.
4. Add the cherries and cook for 1 additional minute.
5. Remove from the heat and cool completely.
6. Cover and refrigerate until ready to use.

Cream Cheese Danish Filling

YIELD: 5 LB 7¼ OZ/2.5 KG

INGREDIENT	US	METRIC
Bread flour	4½ oz	130 g
Salt	¾ oz	2 g
Clear gel starch	2¼ oz	65 g
Butter	7 oz	205 g
Sugar	1 lb	455 g
Vanilla extract	½ tsp	2 g
Eggs	8½ oz	240 g
Cream cheese	3 lb	1.4 kg

1. Sift together the flour, salt, and starch. Set aside.
2. In the bowl of a mixer, cream the butter, sugar, and vanilla until smooth, scraping as needed.
3. Add the sifted dry ingredients to the mixer and mix until combined, scraping as needed.
4. Gradually add the eggs, mixing until smooth. Scrape as needed.
5. With the mixer on medium speed, slowly add the cream cheese in pieces until it is combined, being careful not to overmix. Scrape frequently, ensuring that no large lumps remain (there may be small lumps).
6. Cover and refrigerate until ready to use.

Hazelnut Danish Filling

YIELD: 2 LB 15 OZ/1.3 KG

INGREDIENT	US	METRIC
Milk	9 oz	250 g
Sugar	10½ oz	300 g
Butter	5¼ oz	150 g
Hazelnuts, roasted and ground	1 lb 2 oz	510 g
Cake crumbs	2 oz	60 g
Vanilla extract	½ oz	10 g
Salt	¼ oz	5 g
Eggs	2½ oz	70 g

1. In a saucepan over moderate heat, bring the milk, sugar, and butter to a boil.
2. Stir in the hazelnuts, cake crumbs, vanilla, and salt. Remove from the heat and allow to cool.
3. Fold in the eggs until well combined. Cover and refrigerate until use.

Sticky Buns

DESIRED DOUGH TEMPERATURE: 78°F/26°C

INGREDIENT	US	METRIC	BAKERS' PERCENTAGE
Sponge			
Bread flour	5 lb 8 oz	2.5 kg	50.0%
Milk	5 lb 1 oz	2.3 kg	46.6%
Yeast, instant dry	3 oz	85 g	1.6%
Final Dough			
Sponge	10 lb 13 oz	4.9 kg	98.2%
Butter, softened	2 lb 3¼ oz	1 kg	20.0%
Malt syrup	1 oz	30 g	0.5%
Bread flour	5 lb 8 oz	2.5 kg	50.0%
Eggs	1 lb 14 oz	850 g	17.0%
Salt	3½ oz	100 g	2.0%
Sugar	3 lb 5 oz	1.5 kg	30.0%
Total Dough Weight	24 lb	10.9 kg	217.7%
Pan Smear for Sticky Buns (page 262)	As needed	As needed	

1. Prepare the sponge the same day as the final dough. Combine the flour, milk, and yeast in a mixer, and mix on low speed until homogenous, about 2 minutes. Increase the speed and mix on high for 2 minutes until a dough forms. Ferment at room temperature for about 30 minutes.
2. To prepare the final dough, combine the sponge, about one third of the butter, the malt syrup, flour, eggs, and salt in a mixer. Mix on low speed until homogenous, about 4 minutes. Increase the speed to high and mix until the dough reaches the intense stage of gluten development, about 2 minutes.
3. Add another third of the butter while mixing on high speed, scraping often. Stream in half the sugar, followed by the remaining butter. Add the remaining sugar and mix until combined, about 2 minutes.
4. Mix on high speed until the dough returns to the intense stage of gluten development, about 2 minutes.

5. Remove the dough from the mixer and bulk-ferment, covered, for about 45 minutes.

6. Divide the dough into 1 lb/450 g pieces and preshape into rounds. Bench-rest the dough for about 15 minutes.

7. Roll the dough into a square that is ¼ in/.6 cm thick. Evenly spread with the pan smear, all the way to the edges. Start from one end of the dough and roll toward the other end.

8. Slice the roll into 1-in/2.5-cm-thick pieces. Spread the bottom of a hotel pan with the pan smear and place the dough cut side down in the pan. Proof for about 1 hour or as needed, depending on the environment.

9. Bake in a 375°F/191°C oven until golden brown, about 45 minutes.

10. Invert the pan of buns onto a sheet tray to cool.

Pan Smear for Sticky Buns

YIELD: 11 LB/5.0 KG

INGREDIENT	US	METRIC
Butter, room temperature	2 lb 2¾ oz	985 g
Brown sugar	4 lb 6½ oz	2 kg
Corn syrup	4 lb	1.8 kg
Vanilla paste	1½ oz	40 g
Bread flour, sifted	5¾ oz	165 g
Salt	1½ oz	40 g

1. In the bowl of a mixer fitted with the paddle attachment, cream the butter and brown sugar until light and fluffy, about 4 minutes. Scrape the bowl as needed.

2. Add the corn syrup and vanilla paste, and mix until combined, about 1 minute.

3. Add the flour and salt, and mix just until combined, about 1 minute. Scrape as needed.

4. Transfer to a covered container and store in the refrigerator until needed. Bring to room temperature about 2 hours before using.

Broa

DESIRED DOUGH TEMPERATURE: 80°F/27°C

INGREDIENT	US	METRIC	BAKERS' PERCENTAGE
Corn Soaker			
Coarse cornmeal	2 lb 15 oz	1.3 kg	100.0%
Sea salt	2½ oz	80 g	6.0%
Water, boiling	6 lb	2.7 kg	200.0%
Olive oil	5½ oz	160 g	12.0%
Final Dough			
Water	4 lb 3 oz	1.9 kg	48.0%
Bread flour	7 lb 5 oz	3.3 kg	83.0%
Coarse cornmeal	1 lb 8 oz	680 g	17.0%
Corn Soaker	9 lb 4 oz	4.2 kg	106.0%
Yeast, instant dry	2¾ oz	80 g	2.0%
Total Dough Weight	22 lb 9½ oz	10.2 kg	256.0%

1. Prepare the corn soaker several hours ahead. Combine the cornmeal, salt, and boiling water in a bowl, and mix until the cornmeal is hydrated. Allow to stand for 30 minutes. Add the olive oil and mix until homogenous. Let the mixture stand for 8 hours at room temperature.
2. To mix the final dough, add the water, flour, cornmeal, soaker, and yeast to a mixer. Mix on low speed until homogenous, about 3 minutes. Increase the speed to high and mix until the dough has reached the improved stage of gluten development, about 3 minutes.
3. Remove the dough from the mixer and bulk-ferment, covered, for about 45 minutes.
4. Fold the dough and then ferment for an additional 40 minutes. If the dough is too wet, give the dough 2 folds at 30-minute intervals during the first hour of fermentation.
5. Divide the dough into 1 lb 10½ oz/750 g pieces and preshape into rounds. Allow the rounds to bench-rest for 10 to 15 minutes.
6. Shape the dough into boules or bâtards, and place each seamside up in a floured banneton. Proof for about 1 hour or as needed, depending on the environment.
7. Score the dough and load into a 450°F/232°C oven. Steam 1 second before loading and 3 seconds after loading. Bake for about 45 minutes, or until golden brown.
8. Place the finished breads on a rack to cool.

Croissants

DESIRED DOUGH TEMPERATURE: 75°F/24°C

INGREDIENT	US	METRIC	BAKERS' PERCENTAGE
Dough			
Yeast, fresh	10-1/5 oz	300 g	6.0%
Milk	6 lb 10 oz	3 kg	60.0%
Flour	11 lb	5 kg	100.0%
Butter	1 lb 9 oz	700 g	14.0%
Sugar	1¼ oz	500 g	10.0%
Salt	5¼ oz	150 g	3.0%
Roll-in Butter			
Butter, pliable	6 lb 3 oz	2.8 kg	56.0%
Total Dough Weight	21 lb 4 oz	9.7 kg	193.0%
Egg wash	As needed	As needed	

1. Prepare and laminate the dough a day before making and baking the croissants. To prepare the dough, in the bowl of a mixer, dissolve the yeast in the milk. Add the flour, butter, sugar, and salt and mix on low speed for 5 minutes, making sure to scrape the bowl as needed.

2. Remove the dough from mixer, cover, and rest in the refrigerator for 10 to 15 minutes.

3. To prepare the roll-in butter, wrap the butter in plastic and use a rolling pin to pound and soften the butter. Fold a piece of parchment paper so that it is about 9 by 9 in/23 by 23 cm. Enclose the butter mixture in the parchment and use a rolling pin to roll the butter to the dimensions of the parchment package. Refrigerate the roll-in butter to keep it cold and pliable.

4. Unwrap the butter and enclose it in the dough by placing it in the center of the dough and folding the 4 corners in to enclose the butter. Cover and rest for 30 minutes in the refrigerator.

5. Roll the dough to a rectangle 12 by 28 in/31 by 71 cm and give the dough a three-fold: fold one end of the dough just past the center, and then fold the other half over the first, like a letter (see page 208). There should be 3 layers visible. Place the dough in the refrigerator to rest for 30 minutes.

6. Roll the dough again to a rectangle 12 by 28 in/ 31 by 71 cm, and give the dough a four-fold: fold one end of the dough a quarter of the way. Bring the other end of the dough three-fourths of the way to meet the first edge to create a seam.

Now fold the entire piece of dough in half, like a book (see page 208). There should be 4 layers visible. Place the dough in the refrigerator to rest for 30 minutes.

7. Roll the dough a third time to a rectangle 12 by 28 in/31 by 71 cm, and give the dough a final three-fold.
8. Now roll the dough to a rectangle 14 by 16 in/36 by 41 cm, and cut into equilateral triangles, 7 by 4 in/18 by 10 cm. Cut a ¼-inch slit in the center of the triangle base (you can also use a croissant cutter).
9. With the tip of the triangle pointing away, grasp the dough and gently separate the base, stretching outward slightly. Bring the base of the triangle over the dough and begin rolling it toward the tip. Place tip-side down on a parchment-lined sheet tray.
10. Lightly flatten the croissants with the palm of the hand and apply a coat of egg wash. Proof the dough for about 1 hour or as needed, depending on the environment.
11. Apply a second coat of egg wash.
12. Bake in a 380°F/193°C oven until golden brown, about 20 minutes. Rotate the trays once during baking.
13. Place the finished croissants on a rack to cool.

Pain au Chocolate

DESIRED DOUGH TEMPERATURE: 75°F/24°C

INGREDIENT	US	METRIC	BAKERS' PERCENTAGE
Dough			
Yeast, fresh	10½ oz	300 g	6.0%
Milk	6 lb 10 oz	3 kg	60.0%
Flour	11 lb	5 kg	100.0%
Butter	1 lb 9 oz	700 g	14.0%
Sugar	1 lb 1½ oz	500 g	10.0%
Salt	5¼ oz	150 g	3.0%
Roll-in Butter			
Butter	6 lb 3 oz	2.8 kg	56.0%
Total Dough Weight	21 lb 4 oz	9.7 kg	193.0%
Chocolate batons	2 lb	900 g	
Egg wash	As needed	As needed	

1. Prepare and laminate the dough a day before making and baking the pain au chocolat. To prepare the dough, in the bowl of a mixer, dissolve the yeast in the milk. Add the flour, butter, sugar, and salt and mix on low speed for 5 minutes, making sure to scrape the bowl as needed.
2. Remove the dough from mixer and divide into 4 equal pieces that are 1 lb 9¼ oz/715 g each. Wrap and place each piece of dough on a sheet tray and rest in the refrigerator for 10 to 15 minutes.
3. To prepare the roll-in butter, wrap the butter in plastic and use a rolling pin to pound and soften the butter. Fold a piece of parchment paper in half. Place one fourth of the butter in the paper and roll the butter until it fits the dimensions of the paper. Repeat with the remaining roll-in butter to make a total of 4 roll-ins.
4. Remove one piece of the dough from the refrigerator and roll to a rectangle 16 by 24 in/40 by 60 cm. Place the roll-in butter on one side of the rectangle and fold over the opposite side, enclosing the butter like a book. Seal the edges of the dough and again roll the dough to a rectangle 16 by 24 in/40 by 60 cm.
5. Perform a three-fold 3 times, resting between each fold, ensuring that the butter and dough remain at the same temperature; rest the dough in the refrigerator between folds, if necessary. Repeat with remaining 3 batches of dough and roll-in butter. Cover and refrigerate overnight.
6. The next day, remove the dough from the refrigerator and allow it to come to room temperature. Roll the dough into a rectangle that is ½ in/1 cm thick.
7. Cut the dough into rectangles measuring 3 by 4 in/7.5 by 10 cm. Place 2 chocolate batons ½ in/13 mm apart on the rectangles. They should be visible, but not hanging far off the edges of the dough.
8. Brush the far ends of the rectangles with egg wash and roll the dough toward the egg wash to enclose the chocolate.
9. Place the dough pieces on parchment-lined sheet trays and brush with egg wash. Proof for 30 minutes to 1½ hours or as needed, depending on the environment.
10. Brush with egg wash a second time after proofing.
11. Place the pain au chocolate into a 380°F/193°C oven and immediately lower the temperature to 345°F/174°C and bake until golden brown, about 20 minutes. Rotate the sheet trays once during baking.
12. Place pain au chocolate on a rack to cool.

La Mouna

DESIRED DOUGH TEMPERATURE: 70°F/21°C

INGREDIENT	US	METRIC	BAKERS' PERCENTAGE
Soaker			
Orange peel, grated	1 lb 2 oz	510 g	39.7%
Grand Marnier	3½ fl oz	100 ml	8.0%
Final Dough			
Yeast, fresh	1¼ oz	35 g	7.9%
Orange flower water	1 lb 5 oz	600 ml	9.5%
Bread flour	2 lb 12½ oz	1.3 kg	100.0%
Eggs	1 lb 11 oz	760 g	60.3%
Pâte Fermentée (page 199)	1 lb 5 oz	600 g	89%
Sugar	1 lb 1½ oz	500 g	39.7%
Sea salt	1¼ oz	35 g	2.9%
Butter, softened	3½ oz	100 g	47.6%
Soaker	4 oz	120 g	47.7%
Total Dough Weight	11 lb 3 oz	5.1 kg	404.6%
Egg wash	As needed	As needed	
Vanilla sugar	As needed	As needed	

1. Prepare the soaker (see page 131) 18 hours before preparing the final dough.
2. To make the final dough, dissolve the yeast in the orange blossom water. Combine in a mixer with one fourth of the flour to make a sponge. Add the remaining flour on top of the mixture and allow it to rest until the flour begins to crack.
3. Add the eggs and pâte fermentée, and mix on low speed until homogenous, about 2 minutes. Stream in the sugar and salt and mix until incorporated. Add the butter in small increments and mix until fully incorporated, about 2 minutes. Add the soaker and mix until combined.
4. Increase the speed to high and mix until the dough has reached the improved stage of gluten development, about 6 minutes.
5. Remove the dough from the mixer and bulk-ferment until doubled in size, about 1 hour. Fold once halfway through fermentation.

6. Divide the dough into 1 lb 5 oz/600 g pieces and preshape into rounds. Bench-rest the dough for about 15 minutes.

7. Shape the dough into rounds and use the end of a rolling pin to make a hole in the center of the rounds. Place the rounds seam side down on parchment-lined sheet trays.

8. Brush the rounds with egg wash and proof for about 1 hour or as needed, depending on the environment.

9. Apply a second coat of egg wash and top with vanilla sugar.

10. Bake at 380°F/193°C, with the vent open, until golden brown, about 45 minutes. Remove from the oven and allow to cool completely on a rack.

Brioche

DESIRED DOUGH TEMPERATURE: 77°F/25°C

INGREDIENT	US	METRIC	BAKERS' PERCENTAGE
Eggs	4 lb 6½ oz	2 kg	40.0%
Milk	3 lb 1½ oz	1.4 kg	28.2%
Malt syrup	1 oz	30 g	0.6%
Bread flour	11 lb	5 kg	100.0%
Yeast, instant dry	2 oz	60 g	1.2%
Salt	4½ oz	130 g	2.6%
Butter, softened	5 lb 12 oz	2.6 kg	52.3%
Sugar	1 lb 10 oz	745 g	14.9%
Total Dough Weight	26 lb 7 oz	11.9 kg	239.8%

1. Prepare the dough a day ahead. Combine the eggs, milk, malt syrup, flour, yeast, salt, and about one third of the butter in a mixer. Mix on low speed until homogenous, about 4 minutes. Increase the speed to high and mix until the dough reaches the intense stage of gluten development, about 4 minutes.
2. Return the mixer to low speed, and add another third of the butter (with the mixer running), scraping the dough as needed.
3. While mixing, stream in the sugar, and then add the remaining butter. Increase the speed to high and mix until the dough returns to the intense stage of gluten development, about 2 minutes.
4. Remove the dough from the mixer, and divide it equally between 2 sheet trays. Wrap the trays with plastic wrap and refrigerate overnight.
5. Divide and shape as needed.

CHEF'S NOTE

For Brioche à Tête, divide the dough into 2 oz/57 g pieces.

To form Brioche à tête, roll each piece of dough so that there is a large bulb on one side, with a slim neck, and a smaller bulb at the end.

Use your thumb to make a hole in the large bulb, and pull the smaller end under and through the hole, before placing in a Brioche à tête mold.

For an alternate shaping, create two rounds of dough, one smaller than the other. Place the larger round in the mold and use your thumb to make an imprint. Place the smaller round inside the imprint.

Brioche à tête made with the two-piece method have a higher likelihood of separating during proofing.

Craquilan

DESIRED DOUGH TEMPERATURE: 77°F/25°C

INGREDIENT	US	METRIC	BAKERS' PERCENTAGE
Brioche Dough (#1)	3 lb 9 oz	1.6 kg	60.0%
Brioche Dough (#2)	5 lb 14½ oz	2.7 kg	100.0%
Sugar cubes	1 lb 7¾ oz	670 g	25.0%
Orange zest, grated	1½ oz	40 g	1.5%
Total Dough Weight	11 lb	5.0 kg	186.5%
Egg wash	As needed	As needed	

1. Prepare the brioche dough a day ahead and retard in the refrigerator overnight. Divide the dough into two batches.
2. Divide brioche dough #1 into 1 oz/28 g pieces and shape into rounds. Place the rounds in the freezer until cold.
3. Place brioche dough #2, the sugar cubes, and orange zest in a mixer and mix with the paddle attachment until combined, about 1 minute.
4. Divide the dough into 2½ oz/71 g pieces. On a lightly floured surface, shape the pieces into rounds. Bench-rest the dough for about 10 minutes.
5. Remove the rounds of dough #1 from the freezer and roll into 4-in/10-cm disks. Brush the disks with egg wash, place over the rounds of dough #2, and place into greased brioche floret cups. Brush with egg wash.
6. Proof the dough for 1½ to 2 hours or as needed, depending on the environment.
7. Apply a second coat of egg wash. Use scissors to score the tops of the craquilans.
8. Load into a 400°F/204C oven and immediately lower the temperature to 375°F/191°C. Bake until golden brown, 15 to 18 minutes. Remove from the oven and allow to cool completely on a rack.

Pain de Mais

DESIRED DOUGH TEMPERATURE: 78°F/26°C

INGREDIENT	US	METRIC	BAKERS' PERCENTAGE
Corn Soaker			
Milk, boiling	8 lb 6 oz	3.8 kg	59.8%
Cornmeal	2 lb 14 oz	1.3 kg	21.2%
Milk, cold	4 lb 6½ oz	2 kg	32.0%
Vegetable oil	11 oz	310 g	4.9%
Honey	1 lb 4¼ oz	575 g	9.0%
Final Dough			
Corn Soaker, room temperature	15 lb 14 oz	7.2 kg	113.0%
Malt syrup	1 oz	30 g	0.5%
Bread flour	9 lb 4 oz	4.2 kg	63.1%
Corn meal	2 lb 3 oz	995 g	15.7%
Yeast, instant dry	1 oz	30 g	0.5%
Salt	6 oz	170 g	2.6%
Corn kernels, room temperature	2 lb 13 oz	1.3 kg	20.1%
Total Dough Weight	30 lb 8 oz	13.8 kg	217.8%
Egg wash	As needed	As needed	

1. Make the corn soaker the day before making the final dough. Combine the boiling milk with the cornmeal in a mixer. Rest for 2 minutes before mixing on high speed for 5 minutes.
2. Add the cold milk in 3 parts, scraping well after each addition. Add the oil and mix until incorporated. Add the honey and mix for about 1 more minute, scraping as needed.
3. Transfer the soaker to 2 hotel pans. Cover and refrigerate overnight.
4. To prepare the final dough, combine the soaker, malt syrup, bread flour, corn meal, yeast, salt, and corn in a mixer. Mix on low speed until homogenous, about 4 minutes. Increase the speed to high and mix until the dough reaches the intense stage of gluten development, about 6 minutes.

5. Remove the dough from the mixer and bulk-ferment, covered, for about 45 minutes.
6. Divide the dough into 1 lb 12 oz/795 g pieces and preshape into rounds. Bench-rest the dough for 15 minutes.
7. Shape the dough into 10-in/25-cm oblong rolls and place seam side down into oiled strap pans. Brush with egg wash. Proof the dough for 45 minutes or as needed, depending on the environment.
8. Apply a second coat of egg wash.
9. Load the bread into a 475°F/246°C convection oven and immediately lower the temperature to 400°F/ 204°C. Bake until golden brown, about 16 minutes. Remove from the oven and allow to cool completely on a rack.

Challah

DESIRED DOUGH TEMPERATURE: 84°F/20°C

INGREDIENT	US	METRIC	BAKERS' PERCENTAGE
Sponge			
Water	1 lb 4¼ oz	580 g	11.6%
Malt syrup	1 oz	30 g	0.5%
Bread flour	1 lb 4 oz	570 g	11.4%
Yeast, instant dry	2½ oz	70 g	1.4%
Final Dough			
Water	1 lb 7¾ oz	670 g	13.5%
Sponge	2 lb 10 oz	1.2 kg	24.8%
Eggs	2 lb 7 oz	1.1 kg	22.4%
Egg yolks	10 oz	280 g	5.7%
Honey	10 oz	280 g	5.7%
Vegetable oil	1 lb	460 g	9.2%
Bread flour	9 lb 11¼ oz	4.4 kg	88.6%
Sugar	1 lb 1½ oz	500 g	10.0%
Salt	3¾ oz	105 g	2.1%
Total Dough Weight	20 lb 1 oz	9.1 kg	182.0%
Egg wash	As needed	As needed	

1. Mix the sponge the same day as the final dough and allow to ferment in a large oiled bowl for 20 to 25 minutes.
2. To prepare the final dough, combine the water, sponge, eggs, egg yolks, honey, oil, flour, sugar, and salt in a mixer and mix on low speed until homogenous, 4 to 6 minutes. Increase the speed to high and mix until the dough has reached the intense stage of gluten development, about 4 minutes.
3. Remove the dough from mixer and bulk-ferment, covered, until it doubles in size, about 30 minutes.
4. The dough can be made either into 6-braid loaves or 4-braid loaves. For 6-braid loaves, divide the dough into 6 lb/2.7 kg presses of 36 pieces each. For 4-braid loaves, divide the dough into 4 lb 6½ oz/2 kg presses of 18 pieces each.
5. Preshape the pieces into tight oblong rolls, ensuring that there are no air bubbles. Immediately roll each

piece of dough into a long, tapered strip. For a 6-braid loaf, roll the dough to 9½ in/24 cm long. For a 4-braid, roll the dough to 13 in/33 cm long.

6. Braid the loaves (see below and page 278) and brush with egg wash. Proof the dough for 40 minutes to 1 hour, depending on the environment.

7. Brush the loaves a second time with egg wash.

8. Bake in a 400°F/205°C oven until golden brown, about 40 minutes.

9. Place the finished breads on racks to cool.

Dust the strands with flour and lay the pieces in order: 1, 2, 3, and 4.

Cross 4 into the second position, and then 2 into the fourth position (Remember: outside to inside, inside to outside.)

Bring 1 into the third position, and then 3 into the first position (outside to inside, inside to outside).

Repeat the first two steps until there is no dough left to braid.

Pinch the ends of the dough together and use the sides of your hand to roll off any excess dough.

Finished four-braid Challah.

Dust the strands with flour and lay the pieces in order: 1, 2, 3, 4, 5, and 6.

Bring 5 to the first position, and then bring 6 to the third position (right strands: inside to outside, outside to middle).

Bring 2 to the sixth position, and then bring 1 to the third position (left strands: inside to outside, outside to middle).

Repeat the first two steps until there is no dough left to braid.

Pinch the ends of the dough together and use the sides of your hand to roll off any excess dough.

Finished six-braid Challah.

Pan de Muerto (Day of the Dead Bread)

DESIRED DOUGH TEMPERATURE: 75°F/24°C

INGREDIENT	US	METRIC	BAKERS' PERCENTAGE
Sponge			
Milk	3 lb 8½ oz	1.6 kg	30.9%
Bread flour	5 lb 8 oz	2.5 kg	50.0%
Yeast, instant dry	2¾ oz	75 g	1.5%
Final Dough			
Sponge	9 lb	4.1 kg	82.4%
Eggs	3 lb 12 oz	1.7 kg	33.8%
Malt syrup	1 oz	30 g	0.5%
Bread flour	5 lb 8 oz	2.5 kg	50.0%
Salt	4 oz	115 g	2.3%
Butter, softened, cubed	5 lb 8 oz	2.5 kg	49.5%
Sugar	2 lb 10 oz	1.2 kg	24.8%
Lemon zest, grated	2¾ oz	75 g	1.5%
Orange zest, grated	8 oz	225 g	4.5%
Vanilla extract	1½ fl oz	45 ml	0.9%
Ground cinnamon	½ oz	10 g	0.2%
Orange blossom water	1 fl oz	30 ml	0.5%
Total Dough Weight	27 lb 10 oz	12.5 kg	250.8%
Dough Bones			
Pan de Muerto dough	5 lb	2.3 kg	
Bread flour	1 lb 8 oz	680 g	
Egg wash	As needed	As needed	
Butter, melted	As needed	As needed	
Vanilla sugar	As needed	As needed	

1. Prepare the sponge the same day as mixing the final dough. Combine the milk, flour, and yeast in a mixer, and mix on low speed until homogenous, about 2 minutes. Increase the speed to high and mix for an additional 1 minute. Ferment for 30 minutes.

2. To prepare the final dough, combine the sponge, eggs, malt syrup, flour, and salt with about one third of the butter in a mixer. Mix on low speed until homogenous, about 4 minutes. Increase the speed to high and mix until the dough reaches the intense stage of gluten development, about 4 minutes.

3. Return the mixer to low speed, and add another third of the butter (with the mixer running), scraping the dough as needed.

4. While mixing, stream in the sugar, and then add the lemon zest, orange zest, vanilla, cinnamon, and orange blossom water in 2 additions. Add the remaining butter. Increase the speed to high and mix until the dough returns to the intense stage of gluten development, about 2 minutes.

5. Remove the dough from the mixer and bulk-ferment for 40 minutes. Fold once halfway through fermentation.

6. To make the dough bones, combine the dough and flour and mix until homogenous. Set aside.

7. Divide the bread dough into 8¾ oz/250 g pieces and shape into boules. Place on a parchment-lined sheet tray. Divide the bone dough into ¾ oz/20 g pieces. There will be 3 bones per loaf.

8. Shape the bones by rolling each dough piece into a 6-in/15-cm log. Reserve one third of the logs for rosettes. Use a bench knife or paring knife to cut the center of each end of the log, so there are two ½-in/1-cm-long pieces at the end of each log. Roll each side up toward the outside of the log to create 2 side-by-side circles. Place 2 bones on each round of dough in a cross.

9. Use the remaining logs to make rosettes by rolling each piece into a spiral. Place one rosette in the center of each round of dough, where the bones overlap.

10. Brush the dough with egg wash and proof for 1 to 1½ hours or as needed, depending on the environment.

11. Load the dough into a 400°F/204°C oven and immediately lower the temperature to 365°F/185°C. Bake until golden brown, about 25 minutes.

12. Allow the breads to cool before brushing with the butter and rolling in vanilla sugar.

Gibassier

DESIRED DOUGH TEMPERATURE: 75°F/24°C

INGREDIENT	US	METRIC	BAKERS' PERCENTAGE
Biga			
Milk	9¼ oz	260 g	9.7%
Eggs	3½ oz	100 g	3.7%
Bread flour	1 lb 5½ oz	580 g	21.4%
Yeast, instant dry	¼ tsp	1 g	—
Final Dough			
Water	8 oz	230 g	8.3%
Biga	1 lb 1 oz	945 g	34.8 %
Eggs	1 lb 8¾ oz	700 g	25.9%
Orange blossom water	4½ fl oz	130 mL	4.8%
Orange zest, grated	½ oz	15 g	0.5%
Olive oil	12½ oz	350 g	12.9%
Malt syrup	½ oz	15 g	0.5%
Bread flour	4 lb 10 oz	2.1 kg	78.6%
Yeast, instant dry	2 oz	55 g	2.0%
Salt	2 oz	55 g	2.0%
Butter	14 oz	400 g	14.7 %
Sugar	1 lb 3 oz	535 g	19.7%
Anise seeds	1 oz	30 g	1.0%
Candied orange peel, chopped	14½ oz	410 g	15.0%
Total Dough Weight	13 lb 3½ oz	6.0 kg	220.7%

Egg wash	As needed	As needed
Clarified butter, melted	As needed	As needed
Vanilla sugar	As needed	As needed

1. Prepare the biga 18 hours before mixing the final dough.
2. To prepare the final dough, combine the water, biga, eggs, orange blossom water, zest, oil, malt syrup, flour, yeast, and salt with about one third of the butter in a mixer. Mix on low speed until homogenous, about 4 minutes. Increase the speed to high and mix until the dough reaches the improved stage of gluten development, about 4 minutes.
3. Return the mixer to low speed, and add another third of the butter (with the mixer running), scraping the dough as needed.
4. While mixing, stream in the sugar, and then add the anise seeds and orange peel in 2 additions. Add the remaining butter. Increase the speed to high and mix until the dough returns to the improved stage of gluten development, about 2 minutes.
5. Remove the dough from the mixer and bulk-ferment, covered, for about 2 hours.
6. Divide the dough into 3 ¼ oz/90 g pieces and preshape into tight rounds. Bench-rest the dough for about 10 minutes.
7. Roll the dough into 5-in/13-cm oblongs. Shape into crescents and place on a parchment-lined sheet tray.
8. Use a paring knife to make 3 evenly spaced scores on the outside of each crescent. Brush with egg wash. Proof the dough for about 15 minutes or as needed, depending on the environment.
9. Apply a second coat of egg wash.
10. Bake in a 365°F/185°C oven until golden brown, 12 to 14 minutes.
11. Brush the warm breads with clarified butter and roll in vanilla sugar. Cool before serving.

Pasqua di Colomba

DESIRED DOUGH TEMPERATURE: 75°F/24°C

INGREDIENT	US	METRIC	BAKERS' PERCENTAGE
Sour Feeding			
Water, 65°F/18°C	8 oz	230 g	4.6%
Bread flour	1 lb 1¾ oz	500 g	10.0%
White Sourdough (page 142)	1 lb 1¾ oz	500 g	10.0%
First Dough			
Water, 65°F/18°C	3 lb 5 oz	1.5 kg	29.3%
Bread flour	8 lb 13 oz	4.0 kg	79.9%
Yeast, instant dry	¾ oz	20 g	0.4%
Sugar	3 lb 5 oz	1.5 kg	29.9%
Butter	3 lb 1½ oz	1.4 kg	27.9%
Sour Feeding	2 lb 10 oz	1.2 kg	24.7%
Egg yolks	2 lb 3¼ oz	1.0 kg	19.9%
Final Dough			
Bread flour	1 lb 1¾ oz	500 g	10.0%
Sugar	1 lb 9¼ oz	720 g	14.4%
Salt	4¼ oz	120 g	2.4%
Malt syrup	2¾ oz	75 g	1.5%
Egg yolks	2 lb 3¼ oz	1 kg	19.9%
Lemon zest purée	13½ oz	380 g	7.6%
First Dough	23 lb 9½ oz	10.7 kg	214.6%
Butter, softened	3 lb 5 oz	1.5 kg	30.1%
Candied orange peel, grated	3 lb 1½ oz	1.4 kg	27.3%
Almond paste, cubed, cold	1 lb 1¾ oz	500 g	10.0%
Total Dough Yield	37 lb 4½ oz	16.9 kg	337.8%
Pasqua di Colomba Topping (page 286)	As needed	As needed	

1. Prepare the sour feeding 12 to 14 hours in advance of making the final dough. Mix the ingredients on low speed for 3 minutes or until homogenous.
2. To mix the first dough, combine the water, flour, yeast, sugar, butter, and sour feeding in a mixer. Mix on low speed for 2 minutes, until homogenous, followed by 5 minutes on high speed.
3. Add the egg yolks in 3 additions on low speed, and then mix for 8 minutes on high speed or until the intense gluten stage is reached. Allow the first dough to ferment for 12 to 14 hours.
4. To mix the final dough, place the flour, sugar, salt, malt syrup, egg yolks, lemon zest, and first dough. Mix on low speed for 4 minutes, followed by 3 minutes on high speed. Add the butter in 3 additions, scraping often. Then mix on high speed for 3 minutes or until the intense gluten stage is reached.
5. Add the orange peel and almond paste, and mix on low speed for 1 minute or until ingredients are well combined. Remove from the mixer and allow to ferment 10 minutes.
6. Give the dough a fold and ferment for an additional 10 minutes.
7. Divide the dough into 8 oz/230 g pieces and preshape into a round. Allow to bench-rest for 10 to 15 minutes.
8. Shape the dough into oblong rolls, making a divot in the center of each piece. Place into lightly oiled dove molds. Proof for 1 hour.
9. Pipe the topping onto the rolls, and ferment for 1 more hour.
10. Load the bread into a convection oven at 400°F/204°C. Immediately lower the temperature to 360°F/182°C, and bake for 25 to 30 minutes, or until bread has an internal temperature of 195°F/91°C.
11. Cool before serving.

Pasqua di Colomba Topping

YIELD: 7 LB 4¼ OZ/3.3 KG

INGREDIENT	US	METRIC	BAKERS' PERCENTAGE
Corn meal	2 lb 3¼ oz	1 kg	96.0%
Pastry flour	1½ oz	40 g	4.0%
Almond flour	14¾ oz	415 g	39.7%
Sugar	2 lb 3¼ oz	1 kg	96.0%
Vegetable oil	1½ oz	40 g	4.0%
Vanilla extract	½ oz	10 g	0.8%
Egg whites	1 lb 11½ oz	780 g	75.0%

1. Mix the corn meal, pastry flour, almond flour, and sugar in a mixer. Add the oil, vanilla, and egg whites and mix until homogenous, scraping often.
2. Set at room temperature, covered, until needed.

Sprouted Grain Sponge

INGREDIENT	US	METRIC	BAKERS' PERCENTAGE
Wheat berries	3 lb 5 oz	1.5 kg	66.0%
9-grain mix	1 lb 11¼ oz	775 g	34.0%
Water, 100°F/38°C	6 lb	2.7 kg	120.0%
Total Dough Weight	11 oz	5 kg	220.0%

1. Combine the wheat berries and 9-grain mix in a bowl. Pour the water over the grains and mix to combine.
2. Place in a covered container and leave at room temperature for 2 days.
3. Place the mixture in a strainer to drain. Rinse with warm water for about 10 minutes, and then drain for about 2 hours.
4. Place the mixture in the bowl of a food processor and pulse until coarsely ground. Divide the sponge into 2 lb 3¼ oz/1.0 kg portions and freeze until needed.
5. Remove from the freezer and thaw a day before mixing the final dough.

7 FLATBREADS AND BREADS FROM AROUND THE WORLD

Flatbreads are one of the oldest and most varied categories of breads. Countries all over the world have their versions of flatbread, often even varying vastly by region within those countries. What makes flatbreads so popular and common? They can be made easily, quickly, and with very little equipment. Not all flatbreads are yeast-risen, so many have very little or no rising or resting time. In fact, yeast-raised flatbreads gained prominence only in the last 60 to 70 years. Because they are so thin, flatbreads have a very fast baking time as well. This makes flatbreads easy to reproduce in a variety of professional kitchens.

This chapter discusses common flatbreads (from matzoh to naan), as well as other types of breads from around the world.

A BRIEF HISTORY OF FLATBREADS

The first "flatbreads" were actually grain cakes, originating over 75,000 years ago. Early humans kneaded grains and water together and cooked the resulting cakes on hot rocks or baked them inside ashes by the fire. The breads could also be refreshed over the fire on long journeys. These breads were conceived well before any understanding of fermentation, making them some of the oldest breads ever prepared. Over time, the breads evolved to be made with milled flour and water (as early as 6,000 years ago). The type of flour, the way they were mixed, the way they were shaped, and the way they were cooked all affected the end result.

While the core ingredients for flatbreads remain the same today, the breads vary greatly around the world. Some countries have access to unique or different flours, such as certain rye flours or durum flour. In some countries, refined sweeteners aren't readily available, so ingredients like honey might have been added instead. And in some places, salt is coarse and flaky rather than fine. The resulting breads we know today are as varied as the ingredients used to make them. Naan, a yeast-raised flatbread that originated in India, is still baked much the way as the early flatbreads were baked. That is, naan is baked on the walls of a fire-heated tandoori oven, which is made of stone. Other breads are hung to dry, creating a crisp, cracker-like texture that lengthens the shelf life, ensuring the product stores well through the winter months.

Some flatbreads play important roles outside those of basic foodstuff. Matzoh is a prime example—an unleavened flatbread that is consumed during Passover for those who practice Judaism. Matzoh is also probably the fastest bread to make; the dough has no bench rest or fermentation, and is baked for 35 seconds in a 500°F/260°C environment. Unleavened bread is also used in Christian and Catholic churches in the West to represent the Eucharist. In many countries, flatbreads serve as a substitute for utensils, used to scoop up foods that are often served family style. In other countries, flatbreads are eaten much like regular loaves, topped or filled with spreads or toppings, and eaten like wraps or sandwiches.

VARIETIES OF FLATBREADS

There are dozens of unique flatbreads; a few common ones and their variations are:

- *Chapati and rotis*: Breads of India that are fried in ghee (clarified butter).
- *Puri*: A bread made with whole-grain flour and fried like chapati.
- *Papadam*: A thin fried flatbread made from lentil flour.
- *Pita (Greece), Pitta (Middle East), or Pide (Turkey)*: A thick, soft flatbread made from a salted yeast-leavened dough. Sometimes butter is incorporated. These are used to hold many items, such as zaziki, gyros, and falafel.
- *Lefse:* From Norway, a thin and crispy pancake-like bread.
- *Tortilla*: Made in Central America, tortillas can be made from corn, whole wheat, or plain flour. They are most commonly eaten with fillings.
- *Schuttlebrot*: Also known as Shaker bread, this bread is made with rye flour, water, yeast, and spices (caraway, fennel, anise, and coriander). It is difficult to roll out because of its high hydration; the dough is divided into portions, placed on wooden boards, and then the wet dough is shaken until it forms a thin, flat bread.
- *Knäckebrod*: A bread over 1000 years old, the Vikings made this bread to take on long voyages. It lasted many months and was nutritious.

Flatbreads. From the top left: Pitas, Paratha, Lavash, Sardinian Flatbread, Carta de Musica.

Storing Flatbreads

Many flatbreads have a longer shelf life than typical loaves of bread. Flatbreads with a crisp texture, like Lavash (page 335) or Grissini (page 311) will often last as long as one week at room temperature, stored in an airtight container. Some flatbreads can be refrigerated to lengthen their shelf life—Pita (page 319) and Tortillas (pages 321 and 322) are often refrigerated and then warmed in the oven or in a pan on the stovetop before serving. Depending on the recipe, they can last in refrigeration up to three weeks. Many flatbreads can be frozen without sacrificing quality. Some can be frozen raw (uncooked), while others are frozen after being fully cooked and are warmed in the oven or on the stovetop after they are thawed. Frozen flatbreads can keep for up to 3 months.

Baguette Brötchen

DESIRED DOUGH TEMPERATURE: 75°F/24°C

INGREDIENT	US	METRIC	BAKERS' PERCENTAGE
Water	7 lb 1 oz	3.2 kg	64.0%
Bread flour	9 lb 14¾ oz	4.5 kg	95.0%
White rye flour, organic	1 lb 1½ oz	500 g	5.0%
Malt syrup	3½ oz	100 g	2.0%
Butter	2 oz	50 g	1.0%
Yeast, instant dry	5¼ oz	150 g	3.0%
Sea salt	3¾ oz	110 g	2.2%
Total Dough Weight	19 lb	8.6 kg	172.2%

1. Combine the water, bread flour, rye flour, malt syrup, and butter in a mixer. Mix on low speed for 1½ minutes, until all ingredients are homogenous.
2. Allow the dough to rest in the mixer to autolyse for 10 to 15 minutes.
3. Add the yeast and mix on low speed for 1½ minutes. Add the salt during the last 30 seconds of mixing. Increase the speed and mix on high speed until the dough has reached the improved stage of gluten development, about 2 minutes.
4. Remove the dough from the mixer, cover, and bulk-ferment until it has doubled in size, about 45 minutes.
5. Divide the dough into 2 oz/60 g pieces. Preshape the dough into rounds and bench-rest the dough for 10 to 15 minutes.
6. Shape the dough into rounds or oblong rolls and place seam side down on a parchment-lined sheet tray. There should be room to fit 5 rows of 3 each (see Note). Proof the dough for about 30 minutes or as needed, depending on the environment.
7. Score and load the rolls into a convection oven at 470°F/243°C, steaming 1 second before loading and 3 seconds after loading. Bake until dark golden brown, about 20 minutes.
8. Place the finished rolls on a rack to cool.

CHEF'S NOTE

These rolls can be left plain or topped with seeds, such as sesame seeds. Add the seeds before the rolls are proofed.

Müsli-Brötchen

DESIRED DOUGH TEMPERATURE: 78°F/26°C

INGREDIENT	US	METRIC	BAKERS' PERCENTAGE
Müsli Soaker			
Cold water	12¼ oz	350 g	45.5%
Raisins	4¼ oz	120 g	15.6%
Rolled oats	2¾ oz	80 g	10.4%
Dried apples, diced	2¾ oz	80 g	10.4%
Dried apricots, diced	2¾ oz	80 g	10.4%
Millet	2¾ oz	80 g	10.4%
Final Dough			
Water	2 lb 1½ oz	950 g	21.0%
Bread flour	4 lb 6½ oz	2 kg	100.0%
Butter	5½ oz	160 g	8.0%
Yeast, instant dry	4¼ oz	120 g	6.0%
Whole milk powder	3½ oz	100 g	5.0%
Eggs	3½ oz	100 g	5.0%
Honey	2 oz	60 g	3.0%
Malt syrup	1½ oz	40 g	2.0%
Sugar	3 oz	85 g	4.0%
Sea salt	1½ oz	40 g	2.0%
Müsli soaker	1 lb 11 oz	770 g	39.0%
Sunflower seeds, toasted	3 oz	85 g	4.0%
Hazelnuts, toasted, chopped	3½ oz	100 g	5.0%
Total Dough Weight	10 lb 2½ oz	4.6 kg	204.0%
Egg wash	As needed	As needed	

1. Prepare the soaker 1 to 2 hours before mixing the final dough. Combine all the ingredients.
2. Drain the soaker and reserve the liquid. Combine the soaker liquid with the water for the dough.
3. To mix the final dough, measure the amount for the dough from the combined liquid and combine with the flour, butter, yeast, milk powder, eggs, honey, and malt syrup in a mixer. Mix on low speed until homogenous, about 3 minutes.
4. Stream in the sugar and the salt. Increase the speed and mix until the dough reaches the improved stage of gluten development, about 2 minutes.
5. Return to low speed and add the soaker, seeds, and nuts, and mix on low speed until the ingredients are incorporated, about 1 minute. Remove the dough from the mixer and bulk-ferment, covered, for about 1 hour. Fold halfway through fermentation.

6. Divide the dough into 3 oz/85 g pieces for rolls and 1 lb 4 oz/570 g pieces for loaves. Preshape into rounds, and bench-rest the dough for 10 to 15 minutes.

7. Shape the loaves into rounds and place on a parchment-lined sheet tray in 5 rows of 3 each. Brush with egg wash. Proof for about 1 hour or as needed, depending on the environment.

8. Apply a second coat of egg wash.

9. Load the dough into a 400°F/204°C oven. Bake until golden brown, about 15 minutes for rolls and 35 minutes for loaves.

10. Place the finished breads on a rack to cool.

Pain de Bordelaise

DESIRED DOUGH TEMPERATURE: 76°F/24°C

INGREDIENT	US	METRIC	BAKERS' PERCENTAGE
Water	6 lb 10 oz	3 kg	60.0%
Bread flour	4 lb 6½ oz	2 kg	40.0%
Whole wheat flour	5 lb 8 oz	2.5 kg	50.0%
Coarse white rye flour	1 lb 1¾ oz	500 g	10.0%
Liquid Levain (page 132)	4 lb 6½ oz	2 kg	40.0%
Sea salt	3½ oz	100 g	2.0%
Total Dough Weight	22 lb 4¼ oz	10.1 kg	202.0%

1. Prepare the dough a day ahead. Combine the water, bread flour, whole wheat flour, coarse rye flour, and liquid levain and mix on low speed until homogenous, about 1½ minutes. Stop the mixer and rest the dough to autolyse for 15 to 20 minutes.
2. Resume mixing on low speed for about 1 minute. Add the salt and mix to combine, about 30 seconds. Increase the speed to high and mix until the dough reaches the intense stage of gluten development, about 3 minutes.
3. Remove the dough from the mixer and bulk-ferment, covered, for about 3 hours. Fold once after the first hour of fermentation.
4. Divide the dough into 2 lb 3¼ oz/1 kg pieces and preshape into rounds. Bench-rest the dough for 10 to 15 minutes.
5. Shape the dough into 10-in/25-cm boules or bâtards and place in bannetons dusted with rice flour. Proof the dough overnight in a retarder set at 34°F/1°C.
6. Score the dough and load into a 470°F/243°C oven. Steam 1 second before loading and 3 seconds after loading. Bake until dark golden brown, 20 to 25 minutes.
7. Place the finished breads on a rack to cool.

English Muffins

DESIRED DOUGH TEMPERATURE: 76°F/24°C

INGREDIENT	US	METRIC	BAKERS' PERCENTAGE
Poolish			
Water	2 lb 10 oz	1.2 kg	24.9%
Bread flour	2 lb 10 oz	1.2 kg	24.9%
Yeast, instant dry	½ tsp	2 g	—
Final Dough			
Butter	6 oz	170 g	3.4%
Bread flour	8 lb 6 oz	3.8 kg	75.1%
Water	5 lb 1 oz	2.3 kg	46.3%
Malt syrup	1 oz	30 g	0.6%
Poolish	5 lb 8 oz	2.5 kg	49.7%
Yeast, instant dry	1½ oz	45 g	0.9%
Salt	4 oz	115 g	2.3%
Sugar	1½ oz	45 g	0.9%
Total Dough Weight	19 lb 12 oz	3.9 kg	96.6%

1. Prepare the poolish 18 hours before mixing the final dough.
2. Make the final dough. In the bowl of a mixer, combine the butter and flour. Mix on low speed to rub the butter into the flour until a coarse mixture forms, about 2 minutes. Add the water, malt syrup, poolish, yeast, salt, and sugar. Mix on low speed until homogenous, about 4 minutes. Increase the speed to high and mix until the dough reaches the improved stage of gluten development, about 2 minutes.
3. Remove the dough from the mixer and bulk-ferment, covered, until doubled in size, about 45 minutes. Fold the dough and ferment for an additional 15 minutes.
4. On a lightly floured surface, roll the dough to a ½-in/1.25-cm thickness and cut with a 4-in/10-cm round cutter. Place the rounds on a wooden board. Proof the dough for about 20 minutes or as needed, depending on the environment.
5. Place the rounds on a greased flat top (or greased pan, or griddle, over medium heat) and cook until golden brown, about 2 minutes on each side. Transfer to a parchment-lined sheet tray.
6. Load the muffins into a 450°F/232°C oven and bake until cooked through, 6 to 7 minutes.
7. Place the muffins on a rack to cool.

Multigrain English Muffins

DESIRED DOUGH TEMPERATURE: 76°F/24°C

INGREDIENT	US	METRIC	BAKERS' PERCENTAGE
Soaker			
Water	2 lb 10 oz	1.2 kg	24.0%
9-grain mix	1 lb 9 oz	715 g	14.3%
Flax seeds	10 oz	285 g	5.7%
Poolish			
Water	2 lb 10 oz	1.2 kg	24.9%
Bread flour	2 lb 10 oz	1.2 kg	24.9%
Yeast, instant dry	½ tsp	1.4 g	—
Final Dough			
Butter	3 oz	85 g	1.7%
Bread flour	7 lb 8 oz	3.4 kg	67.1%
Whole wheat flour	14 oz	400 g	8.0%
Water	4 lb 13½ oz	2.2 kg	44.0%
Honey	3 oz	85 g	1.7%
Malt syrup	1 oz	30 g	0.6%
Poolish	5 lb 8 oz	2.5 kg	49.7%
Yeast, instant dry	2 oz	55 g	1.1%
Salt	4½ oz	130 g	2.6%
Soaker	4 lb 13½ oz	2.2 kg	44.0%
Total Dough Weight	24 lb 5 oz	11.03 kg	220.6%

1. Prepare the soaker and poolish 18 hours before mixing the final dough.
2. To prepare the final dough, combine the butter, bread flour, and whole wheat flour in a mixer and mix on low speed to rub the butter into the flour. Mix until the dough is sandy, about 2 minutes.
3. Add the water, honey, malt syrup, poolish, yeast, and salt. Mix on low speed until the dough is homogenous, about 4 minutes. Increase the speed to high and mix for 1 additional minute.
4. Reduce the speed to low again, and add the soaker. Mix for 1 minute, then increase the speed to high and mix until the dough reaches the improved stage of gluten development, about 1 minute.
5. Remove the dough from the mixer and bulk-ferment, covered, until the dough has doubled in size, about 45 minutes. Fold the dough once and ferment for an additional 15 minutes.

6. On a lightly floured surface, roll the dough to a ½-in/1.25-cm thickness and cut with a 4-in/10-cm round cutter. Place the rounds on a wooden board. Proof the dough for about 20 minutes or as needed, depending on the environment.
7. Place the rounds on a greased flat top (or greased pan over medium heat) and cook until golden brown, about 2 minutes on each side. Transfer to a parchment-lined sheet tray.
8. Load the muffins into a 450°F/232°C oven and bake until cooked through, 6 to 7 minutes.
9. Place the muffins on a rack to cool.

Pretzels

DESIRED DOUGH TEMPERATURE: 78°F/26°C

INGREDIENT	US	METRIC	BAKERS' PERCENTAGE
Pâte Fermentée (page 199)	1 lb 3¼ oz	545 g	20.0%
Malt syrup	2½ oz	70 g	2.6%
Bread flour	6 lb	2.7 kg	100.0%
Yeast, instant dry	1½ oz	45 g	1.7%
Salt	1¾ oz	50 g	1.9%
Butter, softened	5 oz	140 g	5.1%
Water	3 lb 2 oz	1.4 kg	52.0%
Total Dough Weight	11 lb	5 kg	183.3%
Dipping Solution			
Water	8 lb	3.6 kg	—
Sodium hydroxide	4½ oz	130 g	—
Coarse pretzel salt	As needed	As needed	

1. Combine the pâte fermentée, malt syrup, flour, yeast, salt, butter, and water and mix on low speed until homogenous, about 4 minutes. Increase the speed and mix on high until the dough reaches the intense stage of gluten development, about 6 minutes.

2. Immediately divide the dough. For large pretzels (5¼ oz/150 g), divide into 6 lb/2.7 kg presses and use a dough divider to make 18 pieces. For small pretzels (2¾ oz/76 g), divide into 6 lb/2.7 kg presses and use a dough divider to make 36 pieces.

3. Preshape the dough into oblong rolls and bench-rest the dough for 10 to 15 minutes.

4. For large pretzels, roll the dough into ropes 32 in/81 cm long. For small pretzels, roll the dough into ropes 24 in/61 cm long. Roll the ropes so that there is a slight hump in the center (see page 303).

5. Hold each rope by both ends and cross the ends over each other. Cross the ends again so that there is a twist in the rope. Bring the 2 ends up to the base of the pretzel and press into the rope on either side of the hump.

6. Place the pretzels on a silicone or parchment-lined sheet tray. (Pretzels can be frozen at this point or refrigerated for use later in the day.)
7. Make the dipping solution by boiling 3 qt/2.88 L water and adding the sodium hydroxide, then add the remaining water to the solution.
8. Wearing protective eye covering and gloves, dip the pretzels into the solution (see Note). Transfer the dipped pretzels to the silicone or parchment-lined baking sheet and sprinkle with some coarse pretzel salt.
9. Load the pretzels into a 475°F/246°C convection oven and immediately lower the temperature to 425°F/218°C. Bake for 12 minutes with the vent half open until deep, golden brown, about 12 minutes. Do not steam.
10. Immediately remove the pretzels from the parchment paper and place on a rack to cool.

CHEF'S NOTE

Take care to keep the dipping solution from touching your skin. If sodium hydroxide solution does come into contact with skin, wash thoroughly with water (without soap). If a rash or blister develops, contact a medical professional.

This recipe can also be used to make pretzel rolls or sticks.

Hold the ends of the tapered strand and cross them over each other.

Bring the ends up to meet the bottom of the pretzel, and press to secure.

Pretzels are dipped in a lye solution to create the signature glossy, brown exterior.

Sprinkle the pretzels with salt immediately after dipping, otherwise it will not adhere to the dough.

Pretzels can be shaped a variety of ways, including sticks and rolls.

Pane di Como

DESIRED DOUGH TEMPERATURE: 80°F/27°C

INGREDIENT	US	METRIC	BAKERS' PERCENTAGE
Sponge			
Water	8¾ oz	250 g	20.0%
Bread flour	2 lb 1 oz	1.3 kg	100.0%
Milk	1 lb 15¼ oz	875 g	70.0%
Malt syrup	2¼ oz	65 g	0.5%
Yeast, instant dry	1¾ tsp	7 g	—
Final Dough			
Water	5 lb 12 oz	2.6 kg	52.0%
Bread flour	11 lb	5 kg	100.0%
Sponge	4 lb 13½ oz	2.2 kg	44.0%
Yeast, instant dry	1¾ oz	50 g	0.1%
Sea salt	4½ oz	125 g	2.5 %
Total Dough Weight	21 lb 15½ oz	10 kg	198.5 %

1. Prepare the sponge the day before mixing the final dough.
2. Prepare the final dough. Combine the water, flour, sponge, and yeast in the bowl of a mixer and mix until homogenous, about 2 minutes. Add the salt and mix until incorporated, about 30 seconds.
3. Increase the speed to high and mix until the dough reaches the improved stage of gluten development, about 3 minutes.
4. Remove the dough from the mixer and bulk-ferment, covered, for about 2¼ hours. Fold once halfway through fermentation.
5. Divide the dough into 1 lb 10 oz/750 g pieces and preshape into rounds. Bench-rest the dough for about 15 minutes.
6. Shape the dough into boules and place seam side up in floured round bannetons. Proof the dough for about 1½ hours or as needed, depending on the environment.
7. Score the dough and load into a 450°F/232°C oven. Steam for 1 second before loading and 3 seconds after loading. Bake until dark golden brown, about 50 minutes.
8. Place the finished breads on a rack to cool.

Pane Pugliese

DESIRED DOUGH TEMPERATURE: 77°F/25°C

INGREDIENT	US	METRIC	BAKERS' PERCENTAGE
Biga, 50% hydration (page 125)	16 lb 8½ oz	7.5 kg	150.0%
Yeast, instant dry	5½ oz	150 g	3.0%
Water	8 lb 13 oz	4 kg	80.0%
Bread flour	11 lb	5 kg	100.0%
Sea salt	7 oz	200 g	4.0%
Total Dough Weight	37 lb 4 oz	16.9 kg	337.0%

1. Prepare the biga the day before mixing the final dough.
2. Prepare the final dough. Dissolve the yeast in about 10 percent of the water and set aside. Combine the remaining water, the biga, and flour in a mixer and mix until homogenous, about 1½ minutes. Add the salt and mix until incorporated, about 30 seconds.
3. Increase the speed to high and mix until the dough reaches the improved stage of gluten development, about 3 minutes.
4. Reduce the speed to low and stream in the yeast mixture. Increase the speed to high and mix until incorporated, about 2 minutes.
5. Remove the dough from the mixer and bulk-ferment, covered, for about 3½ hours. Fold once halfway through fermentation.
6. Divide the dough into 1 lb 5 oz/600 g pieces and preshape into rounds. Bench-rest the dough for about 15 minutes.
7. Shape the dough into bâtards. Place seamside up on a well-floured couche. Proof the dough for about 1½ hours or as needed, depending on the environment.
8. Score the dough and load into a 480°F/249°C oven. Steam for 1 second before loading and 3 seconds after loading. Bake until golden brown, about 40 minutes.
9. Place the finished breads on a rack to cool.

Durum Sour Ciabatta

DESIRED DOUGH TEMPERATURE: 76°F/24°C

INGREDIENT	US	METRIC	BAKERS' PERCENTAGE
Durum Sour (page 204)	4 lb	1.8 kg	35.0%
Water	9 lb 11¼ oz	4.4 kg	87.5%
Bread flour	5 lb 8 oz	2.5 kg	50.0%
Durum flour	5 lb 8 oz	2.5 kg	50.0%
Yeast, instant dry	1 oz	30 g	0.6%
Salt	5½ oz	160 g	3.2%
Total Dough Weight	25 lb 1¾ oz	11.4 kg	226.3%

1. In a plastic tub, dissolve the durum sour in the water. Add the bread flour, durum flour, yeast, and salt, and mix by hand until homogenous. There should be no lumps of flour.
2. Allow dough to bulk-ferment for 90 minutes, giving the dough a fold every 20 minutes during the first hour of fermentation.
3. Place the dough on a well-floured surface and gently stretch it into a 22-in/56-cm square. Use a bench knife to quickly cut the dough into 16 equal pieces. Dust the dough with flour as it is cut to prevent sticking.
4. Transfer the dough pieces to a well-floured couche. Stretch the dough pieces to about 14 in/36 cm as they are laid onto the couche. Proof the dough for 30 minutes or as needed, depending on the environment.
5. Load the dough pieces into a 470°F/243°C oven and immediately lower the temperature to 460°F/238°C. Bake until lightly golden, about 20 minutes, followed by an additional 10 minutes with the vent open.
6. Place the finished breads on a rack to cool.

Whole Wheat Ciabatta

DESIRED DOUGH TEMPERATURE: 76°F/24°C

INGREDIENT	US	METRIC	BAKERS' PERCENTAGE
Biga			
Water	2 lb 3¼ oz	1 kg	18.1%
Bread flour	2 lb 10 oz	1.2 kg	22.0%
Crushed wheat	1 lb 5½ oz	610 g	10.8%
Yeast, instant dry	Pinch	Pinch	—
Final Dough			
Biga	6 lb 6 oz	2.9 kg	50.9%
Water	8 lb 9½ oz	3.9 kg	69.2%
Malt syrup	1½ oz	45 g	0.8%
Bread flour	5 lb 8 oz	2.5 kg	44.1%
Whole wheat flour	2 lb 14 oz	1.3 kg	23.1%
Yeast, instant dry	1 oz	30 g	0.5%
Salt	5 oz	140 g	2.5%
Total Dough Weight	23 lb 10 oz	10.7 kg	191.0%

1. Prepare the biga 18 hours before mixing the final dough.
2. Make the final dough. In a plastic tub, combine the biga with about 80 percent of the water and the malt syrup. Mix by hand to dissolve the biga.
3. Add the bread flour and whole wheat flour along with the yeast and salt. Mix by hand until the ingredients are combined and there are no lumps in the dough. Gradually incorporate the remaining water, breaking down the flour to make a smooth, lump-free dough.
4. Transfer the dough into an oiled container (there should be no more than 20 lb/9 kg of dough per tub). Bulk-ferment the dough, covered, for 2 hours. Fold the dough every 20 minutes during the first hour of fermentation.
5. Place the dough on a well-floured surface and gently stretch it into a 22-in/56-cm square. Use a bench knife to quickly cut the dough into 16 equal pieces. Dust the dough with flour as it is cut, to prevent sticking.
6. Transfer the dough pieces to a well-floured couche. Stretch each piece of dough to about 14 in/36 cm as it is laid onto the couche. Proof the dough for 30 minutes or as needed, depending on the environment.
7. Load the dough into a 470°F/243°C oven and immediately lower the temperature to 460°F/238°C. Bake until lightly golden, about 20 minutes, followed by an additional 10 minutes with the vent open.
8. Place the finished breads on a rack to cool.

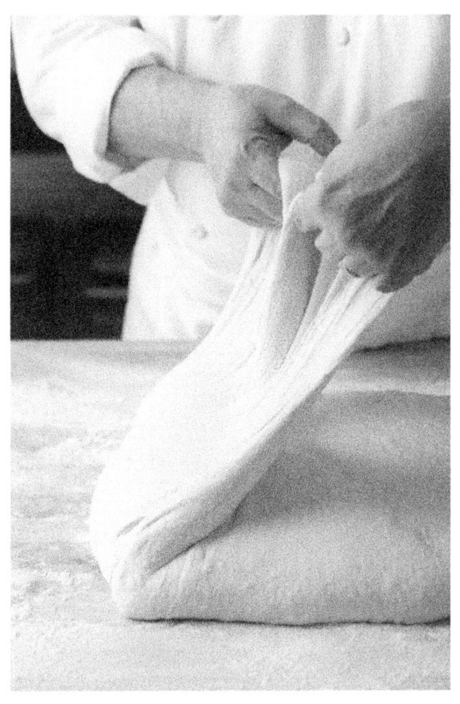

Ciabatta dough is very elastic and requires care when folding.

Divide ciabatta quickly to prevent the dough from quickly deflating.

Carefully slide the bread onto the loader.

Finished ciabatta should have a very open, uniform crumb structure.

Grissini

DESIRED DOUGH TEMPERATURE: 77°F/25°C

INGREDIENT	US	METRIC	BAKERS' PERCENTAGE
Biga, 85% hydration (page 125)	7 oz	200 g	20.0%
Water	1 lb 1 oz	480 g	48.0%
Bread flour	2 lb 3¼ oz	1 kg	100.0%
Yeast, fresh	¾ oz	20 g	2.0%
Sea salt	¾ oz	20 g	2.2%
Butter	3½ oz	100 g	10.0%
Extra-virgin olive oil	3½ oz	100 g	10.0%
Total Dough Weight	4 lb 3 oz	1.9 kg	192.2%
Poppy seeds, caraway seeds, or coarse salt	As needed	As needed	

1. Prepare the biga the day before mixing the final dough.
2. Combine the water, flour, and yeast in a mixer and mix on low speed until homogenous, about 1 minute. Add the salt and the biga, and mix for an additional 30 seconds to incorporate.
3. Increase the speed to high and mix until the dough reaches the intense stage of gluten development, about 3 minutes. Return the mixer to low speed and add the butter and oil. Mix until homogenous, about 3 minutes.
4. Transfer the dough to an oiled container. Cover and bulk-ferment overnight.
5. Divide the dough into 4 equal parts, working with one at a time. Roll each piece into a rectangle ¼ in/6 mm thick. With a pastry wheel or knife, cut ¼-in/6-mm strips of dough.
6. Place the strips on a parchment-lined sheet tray. Brush the dough with water and sprinkle with the seeds or salt. Bake in a 380°F/193°C oven until lightly golden brown and dried, 25 to 30 minutes.
7. Cool on the sheet tray.

Add the olives and herbs at the end of mixing and mix just to combine.

Use a pastry wheel or paring knife to score the dough 7 times.

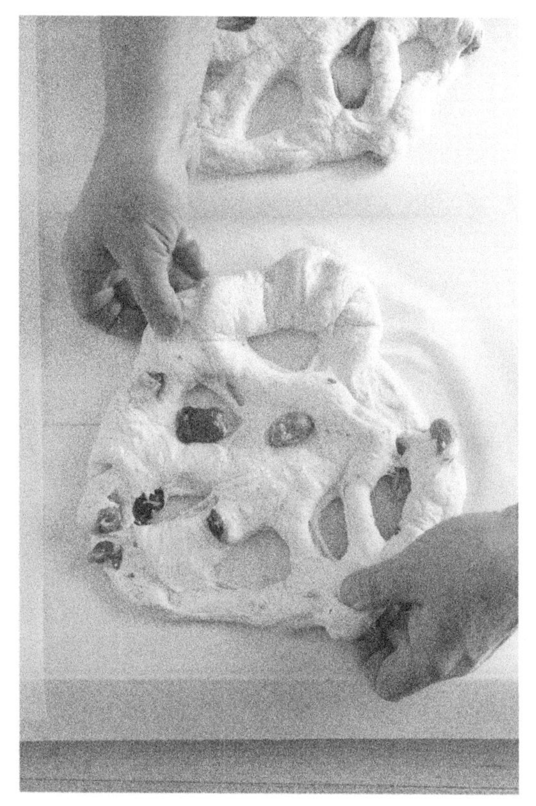

Stretch the dough to open the scores.

Finished Fougasse aux Olives should be golden brown with many visible olives.

Fougasse aux Olives

DESIRED DOUGH TEMPERATURE: 78°F/26°C

INGREDIENT	US	METRIC	BAKERS' PERCENTAGE
Water	7 lb 8 oz	3.4 kg	68.0%
Bread flour	9 lb 14¾ oz	4.5 kg	90.0%
Medium whole rye flour	1 lb 1½ oz	500 g	10.0%
Liquid Levain (page 132)	2 lb 3¼ oz	1 kg	20.0%
Yeast, instant dry	½ oz	15 g	0.2%
Sea salt	3½ oz	100 g	2.0%
Olive oil	3½ oz	100 g	2.0%
Cured Nyon olives	2 lb 12 oz	1.3 kg	25.0%
Thyme, chopped	As needed	As needed	
Total Dough Weight	23 lb 15 oz	10.9 kg	217.2%

1. Combine the water, flours, and liquid levain in a mixer, and mix on low speed until homogenous, about 1½ minutes. Sprinkle the yeast on top of the dough and rest the dough in the mixer for 10 to 15 minutes to autolyse.
2. Resume mixing on low speed to incorporate the yeast, for about 1 minute. Add the salt and mix until combined, about 30 seconds. Increase the speed to high and mix until the dough reaches the intense stage of gluten development, about 3 minutes.
3. Return the mixer to low and, while mixing, stream in the olive oil. Add the olives and thyme, and mix until the ingredients are combined, about 2 minutes.
4. Remove the dough from the mixer and bulk-ferment, covered, until doubled in size, about 3 hours. Fold once after the first hour of fermentation.
5. Divide the dough into 1 lb/450 g pieces and pre-shape into rounds. Bench-rest the dough for about 15 minutes.
6. Shape the dough into rounds and place seam side down onto floured couches. Proof the dough for about 2 hours or as needed, depending on the environment.
7. Stretch the dough rounds into 10-in/25-cm ovals. Using a bench knife or pastry wheel, score 3 times up each side and once in the top center of each dough round. Stretch the ovals to open up the scores.
8. Load the dough into a 470°F/243°C oven. Steam for 1 second before loading and 3 seconds after loading. Bake until dark golden brown, about 25 minutes.
9. Remove the loaves from the oven and allow to cool completely on a rack.

Panini

DESIRED DOUGH TEMPERATURE: 77°F/25°C

INGREDIENT	US	METRIC	BAKERS' PERCENTAGE
Water	6 lb 10 oz	3 kg	58.9%
Bread flour	9 lb 11¼ oz	4.4 kg	87.4%
Durum flour	1 lb 6¼ oz	630 g	12.6%
Salt	4 oz	115 g	2.3%
Sugar	4½ oz	130 g	2.6%
Yeast, instant dry	1 oz	30 g	0.6%
Olive oil, plus for brushing	15¾ oz	445 g	8.9%
Total Dough Weight	19 lb 1½ oz	5.7 kg	173.1%

1. Make the dough a day ahead. Combine the water, bread flour, durum flour, salt, sugar, yeast, and oil in a mixer and mix until homogenous, about 4 minutes. Increase the speed to high and mix until the dough reaches the improved stage of gluten development, about 4 minutes.
2. Remove the dough from the mixer and bulk-ferment, covered, for about 45 minutes.
3. Divide the dough into 10 oz/285 g pieces and pre-shape into rounds. Place onto oiled plastic trays. Brush the dough with oil and cover. Retard the dough in the refrigerator overnight.
4. Bring the dough to room temperature for about 30 minutes before shaping. Roll the dough into 15 by 7-in/38 by 18-cm pieces and place on lightly floured boards. Cover to prevent a skin from forming. Proof for about 20 minutes or as needed, depending on the environment.
5. Transfer the dough to parchment-lined sheet trays and bake in a 375°F/191°C convection oven until golden brown, about 30 minutes.
6. Place the finished breads on a rack to cool.

Naan

DESIRED DOUGH TEMPERATURE: 78°F/26°C

INGREDIENT	US	METRIC	BAKERS' PERCENTAGE
Biga			
Water	1 lb 4 oz	560 g	11.2%
Bread flour	1 lb 14½ oz	865 g	17.3%
Yeast, instant dry	⅛ tsp	0.5 g	—
Final Dough			
Water	3 lb 12 oz	1.7 kg	34.3%
Bread flour	8 lb 9½ oz	3.9 kg	77.7%
Whole wheat flour	8¾ oz	250 g	5.0%
Plain yogurt	4 lb 10 oz	2.1 kg	41.4%
Olive oil	5½ oz	155 g	3.1%
Malt syrup	1¼ oz	35 g	0.7%
Yeast, instant dry	¾ oz	20 g	0.4%
Salt	3½ oz	95 g	1.9%
Biga	3 lb 1½ oz	1.4 kg	28.6%
Total Dough Weight	21 lb 5 oz	9.7 kg	193.2%
Sea salt, garlic, herbs, and spices (optional)	As needed	As needed	
Ghee (clarified butter)	As needed	As needed	

1. Mix the biga 18 hours before mixing the final dough.
2. To prepare the final dough, combine the water, bread flour, whole wheat flour, yogurt, oil, malt syrup, yeast, salt, and biga in a mixer and mix until homogenous, about 4 minutes.
3. Increase the speed to high and mix until the dough reaches the improved stage of gluten development, about 2 minutes.
4. Remove the dough from the mixer and bulk-ferment, covered, for about 1 hour. Fold the dough and ferment an additional 30 minutes.
5. Transfer the dough to a well-floured table and cut into 6 equal pieces. Stretch the dough to 20-in/50-cm-long ovals and place on a well-floured couche. Proof the dough for about 45 minutes or as needed, depending on the environment.

6. Cut each piece of dough into 6 more equal portions and stretch each until 10 in/25 cm long and 3 in/8 cm wide. Lightly depress the centers of the dough pieces with your fingers.
7. Spray the dough with water and garnish with salt, garlic, herbs or spices, if using (see Note).
8. Bake in a 500°F/260°C oven until puffed up, but still light in color, 8 to 10 minutes.
9. Remove from the oven and brush with ghee. Serve immediately.

CHEF'S NOTE

Some garnishes, like some spices, may burn in a hot oven. In this case, add the garnish after baking.

Paratha

DESIRED DOUGH TEMPERATURE: 76°F/24°C

INGREDIENT	US	METRIC	BAKERS' PERCENTAGE
Water	6 lb 14 oz	3.1 kg	62.5%
Whole wheat flour	8 lb 6 oz	3.8 kg	75.0%
Bread flour	2 lb 12 oz	1.3 kg	25.0%
Ghee (clarified butter)	1 lb	450 g	9.0%
Salt	5¼ oz	150 g	3.0%
Total Dough Weight	19 lb 4 oz	8.7 kg	174.5%
Ghee (clarified butter), for brushing	As needed	As needed	

1. Combine the water, whole wheat flour, bread flour, ghee, and salt in a mixer, and mix on low speed until the dough reaches the improved stage of gluten development, about 6 minutes.
2. Remove the dough from the mixer and bulk-ferment, covered, for about 30 minutes.
3. Divide the dough into 1 oz/30 g pieces and preshape into tight rounds, making sure to keep covered with a moist cloth and plastic as you work. Bench-rest the dough for 15 minutes.
4. Roll the rounds to 5-in/13-cm circles and brush with ghee. Roll the circles into balls and rest for 15 minutes.
5. Roll the dough balls into 6-in/15-cm circles.
6. Heat a dry, flat skillet over high heat and cook the paratha until no longer doughy, about 1 minute on each side. Remove from the heat, brush with ghee, and cook until golden brown, about 1 minute on each side.

CHEF'S NOTE

Paratha is an Indian flatbread that is similar to Naan (page 315), but is often stuffed with potatoes, cheese, or vegetables. It is commonly served with chutneys and pickles.

Pita

DESIRED DOUGH TEMPERATURE: 78°F/26°C

INGREDIENT	US	METRIC	BAKERS' PERCENTAGE
Poolish			
Water	12 oz	340 g	23.1%
Bread flour	12 oz	340 g	23.1%
Yeast, instant dry	⅛ tsp	0.5 g	—
Final Dough			
Water	1 lb 8 oz	680 g	46.2%
Bread flour	2 lb 7 oz	1.1 kg	76.9%
Vegetable oil	3 oz	85 g	5.5%
Yeast, instant dry	¼ oz	5 g	0.3%
Malt syrup	¼ oz	5 g	0.6%
Salt	1 oz	30 g	1.5%
Poolish	1 lb 8 oz	680 g	46.2%
Total Dough Weight	5 lb 12 oz	2.6 kg	177.3%

1. Prepare the poolish 18 hours before mixing the final dough.
2. To mix the final dough, combine the water, flour, oil, yeast, malt syrup, salt, and poolish in a mixer. Mix on low speed until homogenous, about 4 minutes. Increase the speed to high and mix until the dough reaches the intense stage of gluten development, about 3 minutes.
3. Remove the dough from the mixer and bulk-ferment, covered, until doubled in size, about 1 hour.
4. Divide the dough into 3 oz/85 g pieces (or 6 lb 12 oz/3 kg presses) and preshape into rounds. Bench-rest the dough for 15 minutes.
5. Roll each piece into a 5 ½-in/14-cm circle and place on a lightly floured couche in 3 rows of 3 each. Proof the dough for about 20 minutes or as needed, depending on the environment.
6. Bake the pitas in a 500°F/260°C oven until puffed, 2½ to 3 minutes.
7. Cool the finished pitas in a cloth to retain moisture.

Whole Wheat Pita

DESIRED DOUGH TEMPERATURE: 78°F/26°C

INGREDIENT	US	METRIC	BAKERS' PERCENTAGE
Poolish			
Water	3 lb 12 oz	1.7 kg	23.1%
Bread flour	2 lb 7 oz	1.1 kg	15.4%
Whole wheat flour	1 lb 3½ oz	555 g	7.7%
Yeast, instant dry	½ tsp	2 g	—
Final Dough			
Water	7 lb 4½ oz	3.3 kg	45.5%
Bread flour	8 lb 9½ oz	3.9 kg	53.8%
Whole wheat flour	3 lb 12 oz	1.7 kg	23.1%
Vegetable oil	14 oz	400 g	5.5%
Yeast, instant dry	¾ oz	20 g	0.3%
Malt syrup	1½ oz	45 g	0.6%
Salt	4 oz	110 g	1.5%
Poolish	7 lb 4½ oz	3.3 kg	46.2%
Total Dough Weight	28 lb 2 oz	12.8 kg	176.6%

1. Mix the poolish 18 hours before preparing the final dough.
2. To prepare the final dough, combine the water, bread flour, whole wheat flour, oil, yeast, malt syrup, salt, and poolish in a mixer. Mix on low speed until homogenous, about 4 minutes. Increase the speed to high and mix until the dough reaches the intense stage of gluten development, about 4 minutes.
3. Remove the dough from the mixer and bulk-ferment, covered, until doubled in size, about 1 hour.
4. Divide the dough into 3 oz/85 g pieces (or into 6 lb 10 oz/3.0 kg presses for 36 pieces on the dough divider). Preshape into rounds and bench-rest the dough for 10 to 15 minutes.
5. Roll each piece into a 5 ½-in/14-cm circle and place on a lightly floured couche, in 3 rows of 3 each. Proof the dough for 20 minutes or as needed, depending on the environment.
6. Bake in a 500°F/260°C hearth oven until the circles are puffed in the center, 2-½ to 3 minutes. They will not develop much color.
7. Place the finished breads in a cloth to cool and retain moisture.

Corn Tortillas

DESIRED DOUGH TEMPERATURE: 78°F/26°C

INGREDIENT	US	METRIC	BAKERS' PERCENTAGE
Water, hot	14 lb 8½ oz	6.6 kg	130.9%
Masa harina	11 lb	5 kg	100.0%
Salt	2¾ oz	75 g	1.5%
Total Dough Weight	25 lb 10½ oz	11.6 kg	232.4%

1. In a large bowl, combine the water, masa harina, and salt. Turn out onto the bench and knead until it forms a homogenous mass, about 2 minutes.
2. Return the dough to the bowl, cover, and rest for 30 minutes.
3. Divide the dough into 1 oz/30 g pieces and preshape into rounds. If using a dough divider, scale 4 lb 8 oz/2.1 kg presses (after dividing on the machine, each resulting piece should then be halved and rounded). Transfer the rounds to a lightly floured plastic sheet tray and rest, covered, for 15 minutes.
4. Line a tortilla press with plastic wrap and lightly grease with pan spray. Flatten each round on the tortilla press.
5. Cook the tortillas in a cast-iron skillet over moderate heat until cooked through and lightly browned.
6. Wrap the tortillas in a towel to retain moisture, or cool and store in a sealed container.

Flour Tortillas

DESIRED DOUGH TEMPERATURE: 78°F/26°C

INGREDIENT	US	METRIC	BAKERS' PERCENTAGE
Bread flour	5 lb 8 oz	2.5 kg	50.0%
Pastry flour	5 lb 8 oz	2.5 kg	50.0%
Shortening	1 lb 9 oz	710 g	14.2%
Salt	4½ oz	125 g	2.5%
Water, hot	5 lb 8 oz	2.5 kg	50.0%
Total Dough Weight	18 lb 5½ oz	8.3 kg	166.7%

1. Combine the bread flour, pastry flour, and shortening in a bowl, and mix to cut the fat in the flour, about 3 minutes. Add the salt and water and mix until it forms a homogenous mass, about 2 minutes.
2. Cover the dough to rest for 30 minutes.
3. Divide the dough into 4 lb/1.8 kg presses and preshape into rounds. Divide on a dough divider into 36 pieces per press (each piece should weigh 1¾ oz/50 g).
4. Place the rounds on a lightly floured plastic sheet tray. Cover and rest the dough for 15 minutes.
5. Roll the dough as thin as possible to a 9-in/23-cm circle (or press the dough with a tortilla press). Stack the tortillas between pieces of parchment paper.
6. Heat a cast-iron skillet, or other flat surface, over high heat and cook the tortillas until they begin to brown, 1 to 2 minutes.
7. Wrap the tortillas in a towel to retain moisture, or cool and store in a sealed container.

Rosette Veneziane

DESIRED DOUGH TEMPERATURE: 80°F/27°C

INGREDIENT	US	METRIC	BAKERS' PERCENTAGE
Biga, 65% hydration (page 125)	3 lb 8½ oz	1.6 kg	80.0%
Water	2 lb 3½ oz	1 kg	52.0%
Bread flour	4 lb 6½ oz	2 kg	100.0%
Yeast, instant dry	2 oz	60 g	3.0%
Butter	8½ oz	240 g	12.0%
Sea salt	1½ oz	40 g	2.0%
Sugar	5½ oz	160 g	8.0%
Olive oil	8½ oz	240 g	12.0%
Total Dough Weight	11 lb 13½ oz	5.4 kg	269.0%
Olive oil	As needed	As needed	

1. Prepare the biga the day before mixing the final dough.
2. Combine the water, biga, flour, yeast, and butter in a mixer and mix until homogenous, about 1½ minutes. Add the salt and sugar and mix until combined, about 1 minute. Increase the speed to high and mix until the dough reaches the improved stage of gluten development, about 2 minutes. Reduce the speed to low and slowly incorporate the olive oil.
3. Remove the dough from the mixer and bulk-ferment, covered, until doubled in size, about 1½ hours. Fold once halfway through fermentation.
4. Divide the dough into 2½ oz/75 g pieces and preshape into rounds. Bench-rest the dough for about 10 minutes.
5. Shape the dough into rounds and brush with olive oil. Stamp each round with a rosette stamper. Proof the dough for about 45 minutes or as needed, depending on the environment.
6. Load the dough into a 420°F/216°C oven. Steam for 1 second before loading and 3 seconds after loading. Bake until golden brown, about 15 minutes.
7. Place the finished breads on a rack to cool.

Focaccia

DESIRED DOUGH TEMPERATURE: 77°F/25°C

INGREDIENT	US	METRIC	BAKERS' PERCENTAGE
Water	4 lb	1.8 kg	67.9%
Bread flour	4 lb 3 oz	1.9 kg	70.9%
Durum flour	1 lb 10½ oz	755 g	29.0%
Yeast, instant dry	3¼ oz	90 g	3.5%
Salt	2 oz	50 g	1.9%
Olive oil	11½ oz	325 g	12.6%
Total Dough Weight	10 lb 10¼ oz	4.8 kg	185.8%

Sea salt, tomatoes, onions	As needed	As needed	
Olive oil, for brushing	As needed	As needed	

1. Combine the water, bread flour, and durum flour in a mixer and mix on low until combined, about 1½ minutes. Sprinkle the yeast on top of the dough and rest the dough in the mixer to autolyse, 10 to 15 minutes.
2. Resume mixing on low speed until the yeast is incorporated, about 1 minute. Add the salt and mix until combined, about 30 seconds. Increase the speed to high and mix until the dough reaches the improved stage of gluten development, about 1 minute.
3. Return the speed to low and stream in the olive oil. Mix until incorporated, about 1 minute.
4. Divide the dough into 2 lb 10 oz/1.2 kg pieces. Lightly shape into rounds and place each piece of dough on a greased, parchment-lined half sheet tray. Proof the dough for 1 to 1½ hours or as needed, depending on the environment. Dimple lightly if the dough is very airy.
5. Stretch the dough into a rectangle roughly 7 by 11 in/18 by 28 cm. Lightly dimple and garnish with salt, tomatoes, or onions as desired.
6. Bake in a 450°F/232°C oven until golden brown, about 30 minutes. Remove the bread from the oven and lightly brush with olive oil.
7. Place the finished bread on a rack to cool.

Focaccia can be hand-mixed to help control the gluten development.

Stretch the focaccia into a rectangle 7 by 11 in/18 by 28 cm on a parchment-lined sheet tray.

After proofing, stipple the dough with your fingers to de-gas the focaccia. This gives the bread its signature appearance.

Focaccia can be topped with any combination of garnishes, such as grapes and thyme, cherry tomatoes, and onions and rosemary.

Pizza

DESIRED DOUGH TEMPERATURE: 77°F/25°C

INGREDIENT	US	METRIC	BAKERS' PERCENTAGE
Bread flour	1 lb 1½ oz	500 g	100.0%
Water	12 oz	340 g	68.0%
Yeast, instant dry	1¼ tsp	5 g	1.0%
Liquid Levain (page 132)	1¾ oz	50 g	10.0%
Salt	½ oz	15 g	2.4%
Olive oil	1 oz	25 g	5.0%
Total Dough Weight	2 lb 1 oz	930 g	186.4%
Rice flour, for dusting	As needed	As needed	
Toppings, such as tomato sauce, cheese, vegetables	As needed	As needed	

1. Place the flour directly on the bench and make a well in the center. Add about one fourth of the water to the well, and place the yeast in the water to dissolve. Draw some flour from the outside of the well to mix with the water to form a paste.
2. Add the liquid levain and mix to combine. Add another fourth of the water and continue mixing. Add the salt, then another fourth of the water, and mix to combine.
3. Add the olive oil, then the remaining water, and mix until the dough becomes a lumpy mass of dough. Knead the dough until it is smooth, about 4 minutes.
4. Divide the dough into 11 oz/310 g pieces. Place on plastic boards until ready to use, 2 to 3 hours or overnight in the refrigerator.
5. Stretch each piece of dough into a 10-in/25-cm circle by hand, or use a rolling pin for a thinner crust. Place on a board dusted with rice flour, ensuring that the dough does not stick and can move easily.
6. Top as desired and load into a 600°F/316°C oven to bake until the crust is golden brown and the topping is properly heated, about 5 minutes.

Rustic Durum Pizza

DESIRED DOUGH TEMPERATURE: 77°F/25°C

INGREDIENT	US	METRIC	BAKERS' PERCENTAGE
Water	7 lb 1 oz	3.2 kg	63.7%
Bread flour	6 lb 6 oz	2.9 kg	57.8%
Durum flour	4 lb 10 oz	2.1 kg	42.2%
Olive oil	6½ oz	185 g	3.7%
Malt syrup	1 oz	30 g	0.5%
Salt	5 oz	135 g	2.7%
Yeast, instant dry	1 oz	30 g	0.5%
Total Dough Weight	18 lb 13½ oz	8.6 kg	171.2%
Olive oil	As needed	As needed	
Rice flour	As needed	As needed	
Topping, such as tomato sauce, cheese, vegetables	As needed	As needed	

1. Make the dough a day ahead. Combine the water, bread flour, durum flour, oil, malt syrup, salt, and yeast in a mixer and mix on low speed until homogenous, about 4 minutes. Increase the speed and mix on high until the dough reaches the improved stage of gluten development, about 3 minutes.
2. Remove the dough from the mixer and bulk-ferment, covered, until doubled in size, about 30 minutes.
3. Divide the dough into 10 oz/285 g pieces and preshape into rounds. Place on an oiled sheet tray and brush with olive oil. Cover and refrigerate overnight (see Note).
4. Remove the dough from the refrigerator 1 hour before shaping and keep at room temperature.
5. Stretch each piece of dough into a 12-in/30-cm circle by hand or use a rolling pin for a thinner crust. Place on a board dusted with rice flour, ensuring that the dough does not stick and can move easily.
6. Brush the outer edges with olive oil and top the pizza as desired.
7. Bake in a 470°F/243°C hearth oven until the crust is golden brown and the topping is properly heated, 10 to 12 minutes.

CHEF'S NOTE

This dough is best when refrigerated overnight, or it can be frozen for later use and thawed in the refrigerator.

Turkish Flatbread

DESIRED DOUGH TEMPERATURE: 78°F/26°C

INGREDIENT	US	METRIC	BAKERS' PERCENTAGE
Water	1 lb 5 oz	600 g	60.0%
Bread flour	2 lb 3¼ oz	1 kg	100.0%
Yeast, instant dry	1¼ oz	35 g	3.5%
Salt	1 oz	30 g	2.5%
Olive oil	1¾ oz	50 g	5.0%
Total Dough Yield	3 lb 12 oz	1.7 kg	171.0%
Egg wash	As needed	As needed	
Caraway seeds	As needed	As needed	

1. Prepare the dough a day ahead. Combine the water, flour, yeast, salt, and oil in a mixer and mix on low speed until combined, about 3 minutes. Increase the speed to high and mix until the dough reaches the intense stage of gluten development, about 2 minutes.
2. Remove the dough from the mixer and bulk-ferment, covered, until the dough doubles in size, about 10 hours.
3. Divide the dough into 1 lb/450 g pieces and pre-shape into rounds. Bench-rest the dough for about 15 minutes.
4. Roll the dough into 8-in/20-cm circles. Transfer to parchment-lined sheet trays and brush with egg wash. Proof for about 1 hour or as needed, depending on the environment.
5. Apply a second coat of egg wash and sprinkle with caraway seeds. Load the dough into a 400°F/204°C oven and bake until golden brown, about 30 minutes.
6. Place the finished breads on a rack to cool.

Kalyra

DESIRED DOUGH TEMPERATURE: 78°F/26°C

INGREDIENT	US	METRIC	BAKERS' PERCENTAGE
Dough			
Almond paste	1 lb 5 oz	595 g	11.9%
Butter, softened	5 lb 4½ oz	2.4 kg	47.6%
Milk	¾ oz	20 g	0.4%
Honey	2 lb 1½ oz	950 g	19.0%
Malt syrup	1¾ oz	50 g	1.0%
Bread flour	6 lb 10 oz	3 kg	60.0%
Pastry flour	4 lb 6½ oz	2 kg	40.0%
Yeast, instant dry	4¼ oz	120 g	2.4%
Salt	6¾ oz	190 g	3.8%
Total Dough Weight	20 lb 9 oz	9.3 kg	67.1 %
Honey Glaze			
Honey	1 lb 1½ oz	500 g	—
Salt	1¼ tsp	5 g	—
Poppy seeds	As needed	As needed	

1. In a tabletop mixer, combine the almond paste and butter. Mix on medium speed until softened, about 2 minutes.
2. Combine the almond paste mixture with the milk, honey, malt syrup, bread flour, pastry flour, yeast, and salt in a mixer. Mix on low speed until homogenous, about 4 minutes. Increase the speed to high and mix until the dough reaches the intense stage of gluten development, about 4 minutes.
3. Remove the dough from the mixer and bulk-ferment, covered, until doubled in size, about 45 minutes.
4. Divide the dough into 5 lb/2.2 kg pieces and pre-shape into tight rounds. Bench-rest the dough for 10 to 15 minutes.

5. Use a dough divider to divide each piece of dough into 36 pieces. Each piece will weigh 2¼ oz/65 g.

6. Roll each piece of dough into a 4½-in/11-cm circle. Use a round cutter to decorate the dough (see Note).

7. Combine the honey and salt in a small saucepan and heat to make the glaze. Brush the dough with the warm glaze and sprinkle with poppy seeds. Proof the dough for 25 minutes or as needed, depending on the environment.

8. Bake the dough in a 350°F/177°C convection oven until golden brown, about 15 minutes.

9. Remove breads from the oven and brush again with the honey glaze. Place the finished breads on a rack to cool.

CHEF'S NOTE

There are many ways to "decorate" the dough. Use small round cutters to cut circles from the interior of the dough or along the edges to create a scalloped edge.

Roasted Garlic and Parmesan Flatbread

DESIRED DOUGH TEMPERATURE: 76°F/24°C

INGREDIENT	US	METRIC	BAKERS' PERCENTAGE
Water	7 lb 15 oz	3.6 kg	37.3%
Bread flour	1 lb ¼ oz	465 g	4.8%
Pastry flour	4 lb 13½ oz	2.2 kg	22.9%
Whole wheat flour	5 lb 1 oz	2.3 kg	24.1%
Semolina flour	10 lb 5¾ oz	4.7 kg	48.2%
Olive oil	2 lb 10 oz	1.2 kg	12.0%
Salt	2½ oz	70 g	0.7%
Malt syrup	1½ oz	50 g	0.5%
Liquid Levain (page 132)	2 lb 10 oz	1.2 kg	12.0%
Garlic, roasted	1 lb 15 oz	870 g	9.0%
Parmesan, finely grated	2 lb 10 oz	1.2 kg	12.0%
Total Dough Weight	39 lb 1½ oz	17.7 kg	183.7%
Olive oil	As needed	As needed	
Sea salt	As needed	As needed	
Parmesan, grated	As needed	As needed	

1. Combine the water, bread flour, pastry flour, whole wheat flour, semolina flour, oil, salt, malt syrup, liquid levain, garlic, and cheese in a mixer and mix on low speed until homogenous, about 4 minutes.
2. Immediately divide the dough into 8 oz/225 g pieces and preshape into rounds. Place on a parchment-lined sheet tray and cover. Bulk-ferment for about 1 hour.
3. Roll the dough through a pasta machine, starting at the highest setting. Roll the dough until you reach setting #2, or about 2 mm (almost paper thin).
4. Transfer the dough to a parchment-lined sheet tray and lightly brush with olive oil. (The dough can be stacked and frozen at this point, if desired.) Bench-rest the dough for 1 hour.
5. Before baking, sprinkle the dough with sea salt and more cheese. Bake in a 350°F/177°C oven with the vent open until golden brown, 12 to 16 minutes.
6. Cool the flatbread completely before breaking into pieces to serve.

Ciriole

DESIRED DOUGH TEMPERATURE: 77°F/25°C

INGREDIENT	US	METRIC	BAKERS' PERCENTAGE
Water	8 lb 7⅔ oz	3.85 kg	
Bread flour	15 lb 7 oz	7 kg	100.0%
Malt syrup	12½ oz	350 g	5.0%
Yeast, instant dry	8½ oz	245 g	3.5%
Sea salt	5 oz	140 g	2.0%
Olive oil	11¼ oz	315 g	4.5%
Total Dough Weight	26 lb 3½ oz	11.9 kg	170.0%

1. Combine the water, flour, and malt syrup in a mixer and mix on low speed until homogenous, about 1½ minutes. Stop the mixer, sprinkle the yeast on top of the dough, and rest to autolyse for 10 to 15 minutes.
2. Resume mixing until the yeast is incorporated, about 1 minute. Add the salt and mix until combined, about 30 seconds. Increase the speed to high and mix until the dough reaches the improved stage of gluten development, about 2½ minutes.
3. Return the mixer to low and stream in the oil. Mix until combined, about 1 minute.
4. Remove the dough from the mixer and bulk-ferment, covered, for about 1 hour. Fold halfway through fermentation.
5. Divide the dough into 3 ½ oz/100 g pieces and preshape into rounds. Bench-rest the dough for about 15 minutes.
6. Shape the dough into small bâtards and place seam side down on a floured couche. Proof for about 45 minutes or as needed, depending on the environment.
7. Lightly dust the dough with flour, score the bâtards, and load into a 470°F/243°C oven. Steam 1 second before loading and 3 seconds after loading. Bake until golden brown, about 25 minutes.
8. Place the finished breads on a rack to cool.

Lavash

DESIRED DOUGH TEMPERATURE: 80°F/27°C

INGREDIENT	US	METRIC	BAKERS' PERCENTAGE
Water	4 lb 6½ oz	2 kg	66.0%
High-gluten flour (13–15% protein)	6 lb 10 oz	3 kg	100.0%
Yeast, compressed	2 oz	60 g	2.4%
Honey	3½ oz	100 g	4.0%
Sea salt	1½ oz	40 g	1.6%
Total Dough Yield	11 lb 7¾ oz	5.2 kg	174.0%
Sesame seeds or za'atar	As needed	As needed	

1. Combine the water, flour, yeast, honey, and salt in a mixer and mix on low speed for 5 minutes. Increase the speed and mix on high speed for an additional minute.
2. Remove the dough from the mixer and immediately divide it into 12¼ oz/350 g pieces. Preshape into logs and bench-rest the dough for 10 to 15 minutes.
3. Roll the dough into rectangles about 7 by 12 in/18 by 30 cm long and stretch until they are very thin (about 2 mm thick). There should be enough room to fit 2 side by side on the back of an oiled sheet tray.
4. Spray the dough with water and garnish with sesame seeds or za'atar, as desired.
5. Bake in a 350°F/177°C convection oven until golden brown and crisp, about 8 minutes.
6. Cool completely before breaking into pieces to serve.

Sardinian Flatbread

DESIRED DOUGH TEMPERATURE: 76°F/24°C

INGREDIENT	US	METRIC	BAKERS' PERCENTAGE
Water	4 lb 6½ oz	2 kg	39.8%
Olive oil	1 lb 4 oz	570 g	11.4%
Malt syrup	1 oz	30 g	0.5%
Bread flour	2 lb 7 oz	1.1 kg	21.6%
Pastry flour	2 lb 7 oz	1.1 kg	21.6%
Semolina flour	6 lb 2½ oz	2.8 kg	56.8%
Salt	1¼ oz	35 g	0.7%
Total Dough Weight	16 lb 12½ oz	7.6 kg	152.3%
Olive oil	As needed	As needed	
Seeds, grated hard cheese, or salt	As needed	As needed	

1. Combine the water, oil, malt syrup, bread flour, pastry flour, semolina flour, and salt in a mixer and mix on low speed until homogenous, about 4 minutes.
2. Immediately divide the dough into 8 oz/225 g pieces and preshape into rounds. Place on a parchment-lined sheet tray and cover. Bulk-ferment for about 1 hour.
3. Roll the dough through a pasta machine, starting at the highest setting. Roll the dough until you reach setting #2, or about 2 mm (almost paper thin).
4. Transfer the dough to a parchment-lined sheet tray and lightly brush with olive oil. (The dough can be stacked and frozen at this point, if desired.) Bench-rest the dough for 1 hour.
5. Before baking, spray the dough with water and garnish as desired. Bake in a 350°F/177°F oven with the vent open until golden brown, 12 to 16 minutes.
6. Cool the bread completely before breaking into pieces to serve.

Sangak

DESIRED DOUGH TEMPERATURE: 78°F/26°C

INGREDIENT	US	METRIC	BAKERS' PERCENTAGE
Water	1 lb 2¼ oz	800 g	44.4%
Olive oil	8 oz	225 g	12.5%
Whole wheat flour	4 lb	1.8 kg	100.0%
Salt	½ oz	14 g	7.7%
Honey	3½ oz	100 g	5.5%
Total Dough Weight	4 lb 4½ oz	2.9 kg	170.1%

1. Combine the water, oil, flour, salt, and honey in a mixer, and mix on low speed until homogenous, about 4 minutes.
2. Immediately divide the dough into 12½ oz/350 g pieces and preshape into rounds. Bench-rest the dough, covered, for 30 minutes.
3. Stretch each piece into a rectangle 18 by 24 in/46 by 61 cm on a parchment-lined sheet tray (see Note).
4. Bake in a 600°F/316°C oven until golden brown, about 6 minutes.

CHEF'S NOTE

This traditional Iranian flatbread is traditionally baked on clean, oiled stones to produce a textured appearance.

Carta da Musica

DESIRED DOUGH TEMPERATURE: 77°F/25°C

INGREDIENT	US	METRIC	BAKERS' PERCENTAGE
Water	1 lb 7½ oz	670 g	67.0%
Fine durum flour	2 lb 3¼ oz	1 kg	100.0%
Sea salt	¼ oz	5 g	0.3%
Yeast, instant dry	⅓ oz	10 g	1.0%
Total Dough Weight	3 lb 11¼ oz	1.7 kg	168.3%

1. Combine the water, durum flour, salt, and yeast in a mixer and mix on low speed until the dough reaches the improved stage of gluten development, about 4 minutes.
2. Divide the dough into 2½ oz/75 g pieces. Preshape into rounds and bench-rest the dough, covered, for about 20 minutes (see Note).
3. Shape the dough into rounds and place on floured boards. Proof for about 1 hour or as needed, depending on the environment.
4. Use a rolling pin to roll the dough into 8-in/20-cm circles. They should be thin and flat.
5. Load the dough into a 500°F/260°C oven and bake, flipping once during baking, until the breads are golden and crisp, about 2 minutes on each side.
6. Place the finished breads on a rack to cool.

CHEF'S NOTE

The shaped dough can be placed on an oiled board and left in a refrigerator to retard overnight.

8 RYE BREADS AND ROLLS

As briefly discussed in Chapter 3, flours milled from rye berries behave differently from wheat flours. While rye flours have distinct colors and unique flavors, they have relatively low levels of protein and produce heavy, dense breads. Many bread bakers do not fully understand how to utilize rye flours to produce high-quality bread. Because of this, rye flour is often mixed with white flour to produce better results. But another excellent way to utilize rye flour is by making sourdough (see page 145). Rye has very high amounts of the amylase enzyme, which is also present in yeast (see page 122). Because of this, rye flours respond well to fermentation, and their flavors only intensify through this process.

A BRIEF HISTORY OF RYE

Rye (*Secale cereale*) is one of the most prominent ingredients in the bread baking of both the Middle East and northern Europe. In Germany especially, where beer and bread reign supreme, Germans have become well known for their understanding of fermentation—malting barley to make beer and souring rye to make bread. Germans eat more bread than most other Europeans and three times as much as most Americans.

Just as Egyptians became known for their eating of bread, Germanic people were often called *Ruggi*, which means "rye eaters." While rye plants evolved later than wheat plants in the Middle East, they had become increasingly common by 3000 B.C.E, particularly in southern Russia, central Europe, and northern Europe, where temperatures were generally cooler and reached more extremes. When a climate is too chilly for wheat, rye can still thrive. When the soil is poor and many crops fail, rye prospers. These attributes made rye a popular crop and, subsequently, a prominent food source. Still to this day, rye is grown mostly on the Northern European Plain, a large stretch of land that runs from the lowlands of Belgium to the Ural Mountains of Russia.

In all these regions, bread is an important food source and a way of life, finding its place in multiple aspects of culture, from religion, to folk tales, to holiday traditions. For more on the history of bread, see Chapter 1.

RYE BASICS

Rye flour is different from wheat flour in many ways. Understanding these differences is the first step toward producing excellent products made from rye flour. The rye grain is not husked before milling; this shortens and eases the milling process, and also retains the great nutritional benefits of rye flour. It is important to know that rye flour has:

- **Fewer gluten-forming proteins:** Rye flour contains less glutenin and gliadin, the proteins that form gluten bonds (see Chapter 4). Therefore, rye doughs have a minimal, generally weak gluten structure. The gluten network's only strength comes from the starches in the flour.
- **Less starch content:** Rye flour has less starch than wheat flour, though not by an incredibly large margin. The primary difference is that the starches in rye flour gelatinize at a lower temperature. While wheat flour gelatinizes at 158°F/70°C, rye flour gelatinizes at 122°F/50°C. In addition, rye flour has a greater concentration of starch-decomposing enzymes, which negatively affect the structure of the starch. The effects of this can deteriorate the crumb structure of the finished loaf and contribute to a bland taste and gummy texture.
- **More soluble sugars:** With greater quantities of soluble sugar than wheat flour, rye flour has higher levels of naturally occurring nutrients, which serve to feed the yeast.
- **Higher levels of enzyme activity:** In addition to the previously mentioned starch-decomposing enzymes, rye flour has high levels of amylase. Amylase, also naturally present in yeast, provides an excellent basis for a sour culture. Rye flour also has a higher percentage of ash and mineral content, which lessen the volume of a finished loaf.
- **More water-binding pentosans (gums):** Rye flour has high quantities of pentosans, or long chains of 5-carbon sugars. These compounds have a high water-binding capacity, which increases the dough's ability to retain moisture (significantly higher than wheat flour). These pentosans take a long time to digest, which makes rye breads more filling than products made with wheat flour.

TYPES OF RYE FLOUR

There are four primary types of rye flour: white, medium, dark, and pumpernickel. Each type has a different flavor, appearance, and makeup—and each produces a very different product. Rye breads are typically defined by how much rye flour they include. Bread labels often say "30 percent dark rye" or "100 percent rye" to indicate the percentage of the total flour that is rye flour. In Europe, there are strict legal definitions for what can be called rye bread. If there is less than 50 percent rye flour in the bread, it must be labeled as "rye mixed." If the bread contains between 50 and 99 percent rye flour, it is "rye bread"; 100 percent rye bread must be made from all rye flour, of course.

The United States does not have these laws, but a good bakery will usually know how much rye flour is in its rye bread. Wheat flour is easier to work with, which is why rye flour is typically mixed with a portion of wheat flour to make certain loaves. Once more than 30 percent rye flour is incorporated into a bread, however, the dough begins to behave like a rye dough and the considerations discussed in this chapter must be taken into account when baking the bread.

White Rye Flour

- *Color/texture:* A shade or two darker than most white wheat flours; finely ground
- *Flavor:* Lightest and most mild, this flour is milled from just the interior of the rye kernel.
- *Protein content:* 8 to 10 percent

Medium Rye Flour

- *Color/texture:* Clearly a rye flour, but still relatively pale in color; finely ground
- *Flavor:* More nutty and flavorful than white rye, milled using the bulk of the kernel's interior (higher bran content than white rye)
- Protein content: 9 to 11 percent

Dark Rye Flour

- Color/texture: Relatively dark brown/gray in color; fine to medium grind
- Flavor: More intensely flavorful, milled using the inner and outer portion of the endosperm (very high protein content)
- *Protein content:* 14 to 17 percent

Pumpernickel Flour

- *Color/texture:* Very dark; coarsely ground (also known as rye meal)
- *Flavor:* The most intensely flavored, milled from the entire rye kernel. Milling the whole kernel raises the protein content, as does using more bran in the milling process.
- *Protein content:* 8 to 12 percent

Storing Rye Breads

Rye bread has different storage concerns than many other varieties of bread. Unlike many varieties of artisanal breads, rye breads do not freeze well. This is due to the high percentage of water in rye doughs, especially in doughs with a high percentage of rye. In addition, it's important to understand that rye flour has more natural oils than other flours, which makes it prone to rancidity. Many rye breads will keep successfully at room temperature for up to a week (which can give it a significantly longer shelf life than the average loaf of bread). However, if kept under refrigeration (wrapped tightly in plastic wrap), loaves can last up to two weeks.

MAKING RYE SOURDOUGH

Rye flour makes excellent sourdough cultures, owing to its high amylase enzyme content. However, just as with any rye loaf, there are some special considerations. Because rye flour absorbs water quickly, the rye sour cultures need salt to help control the speed of fermentation.

The sourdough process is probably the oldest method used in the preparation of breads from cereals. It was developed through the observation that, after some time, a mixture of flour and water shows a change that is now called fermentation. After more in-depth investigations, people learned that these changes are a result of activities from yeast and bacteria. After they observed how these microorganisms work, this knowledge was put to use in the production of sourdough breads. All living organisms need food, water, oxygen, and the right temperature, the latter being the most important.

The dough supplies the necessary food and nutrients to the yeast, which include water-soluble carbohydrates, starches that are highly present in the flour, soluble albumins, and traces of nutrient salt. The necessary water is provided through hydration of the dough. Oxygen is provided by sifting the flour and mixing the sour. The baker must control the proper dough temperature; yeast cells sufficiently multiply at temperatures of 75 to 77°F/24 to 25°C. Using a higher temperature will cause more carbon dioxide to develop, aiding in the development of the dough. The acid bacteria, foremost being lactic bacteria, find nutrients in the water-flour mixture, but they require a warmer temperature than yeast. For the best development, a temperature of 95°F/35°C is ideal.

The Three-Stage Rye Sourdough Method

The different requirements of the microorganisms in a sour culture can be satisfied only through time, the rate of hydration, and temperature. Ultimately, following the correct feeding procedure and meeting the requirements of the different microorganisms leads to production of the multi-stage sourdough. The baker is responsible for adapting to the needs of the microorganisms by changing the ratio of flour, water, and temperature.

The first step in the three-step rye sourdough method is called the *refresher sponge*, which multiplies the yeast cells (see page 377). It is necessary for the dough temperature to be 75 to 77°F/24 to 25°C. Also, a soft dough consistency is preferable because the yeast cells reproduce more efficiently in softer doughs.

The next step is the *base sour*, which promotes the development of aroma. Because the base sour is stiff, it develops acetic acid and very little yeast. Depending on the baker's work flow, this step is done typically overnight, so the base sour will be a higher temperature at the end of fermentation than at the beginning. It is important to choose a lower water temperature for good development of the acid bacteria. Water that is too warm can damage the yeast and coagulate the starches in the flour.

The goal is to create favorable conditions for the development of acid bacteria, but not to hurt the yeast and flour. The average temperature of the base sour should be around 86°F/30°C. During the first two steps, more yeast cells and acetic bacteria are developed. After that, the primary goal is to change the yeast from developing yeast cells to leavening the bread. The base sour, thus, is used to make the *full sour*, the third step in the three-step rye sourdough method.

The third step supports the fermenting ability of the yeast in the dough to develop aroma and moisture-retaining properties. The softer consistency and higher temperature (89°F/32°C) of the full sour promotes the development of lactic acid. The baker not only has to recognize the changes in the development of the yeast and acid bacteria, but also see the change in the flour. Necessary adjustments, then, must be made to the flour.

Principles of Rye Sourdough Management

The main components in flour are starches, soluble sugars, proteins, and minerals; enzymes are present only in small quantities. What happens with these components of the flour? The albumin is already largely available in soluble form; this solubility increases during the sourdough process. The rye protein does not change to gluten, so significant bread structure cannot be expected from the rye proteins.

The changes in the starches are more important. First, the starches increase before and after the steps in the sourdough method; there is rehydration of the starch molecules. The starch gains the ability to absorb water and swell, but this does not happen in rye dough without a sour. The enzymes initiate the breakdown of sugar to fructose and glucose; however, the enzymes are sensitive to acid and gradually cease activity and ultimately die. This occurs at a pH level of 7. Thus, it is possible to regulate the speed of the degradation of the starches.

Bakers desire, to a certain degree, the degradation of the starches because they provide the necessary nutrients for the yeast; however, the goal is not to reduce the dough-forming ability of the starches. If the enzymatic activity is too strong, a sufficient amount of simple sugars will be produced in a short time. Thus, it is necessary to reduce the enzymatic activity by controlling the acid development in the sourdough through adjustment of the temperature and hydration.

In the rye sourdough process, it is imperative to work with exact temperatures and hydration. These two factors are crucial for the development and activity of the microorganisms, as well as for changes in the flour. This specifically pertains to the starches, which are crucial for proper development of the crust and crumb of a rye bread. Therefore, it is important to use 40 to 50 percent of the total rye flour in the sourdough. The starches in the rye flour undergo a transformation, changing the characteristics and readiness to gelatinize. Also, it is important that the starches gelatinize when the bread has reached its full volume.

While a baker can assess the fermentation of the sourdough by touch, visual examination includes noting changes in the flour. The baker can primarily analyze the acid level, but temperature measurements should also be made, not only at the start of each step but also at the end of each step. Only a sourdough that shows warming during fermentation will be well developed.

100% Rye Bread

DESIRED DOUGH TEMPERATURE: 80°F/27°C

INGREDIENT	US	METRIC	BAKERS' PERCENTAGE
Water	7 lb 11 oz	3.5 kg	70.6%
Medium rye flour	11 lb	5 kg	100.0%
Yeast, instant dry	1½ oz	45 g	0.9%
Salt	4¼ oz	120 g	2.4%
Rye Sour (page 147)	10 lb 2½ oz	4.6 kg	91.8%
Total Dough Weight	29 lb 4½ oz	13.3 kg	265.7%
Rolled oats	As needed	As needed	
Vegetable oil	As needed	As needed	

1. Combine the water, flour, yeast, salt, and rye sour in a mixer and mix on low speed until a homogenous dough forms, about 8 minutes. The dough should have a plastic consistency, rather than be elastic.
2. Remove the dough from mixer and bulk-ferment, covered, for about 15 minutes.
3. Scale into 1 lb 8 oz/680 g pieces and shape into rounds. Moisten the dough by rolling it on a damp linen-lined sheet tray, then roll in the oats.
4. Place 2 loaves in each loaf pan, brushing with oil between the loaves.
5. Load the dough into a 470°F/243°C oven, and immediately drop the temperature to 440°F/227°C. Bake until the bread is golden brown, 45 to 60 minutes.
6. Remove from oven and immediately take out of the loaf pans. Place the finished breads on a rack to cool.

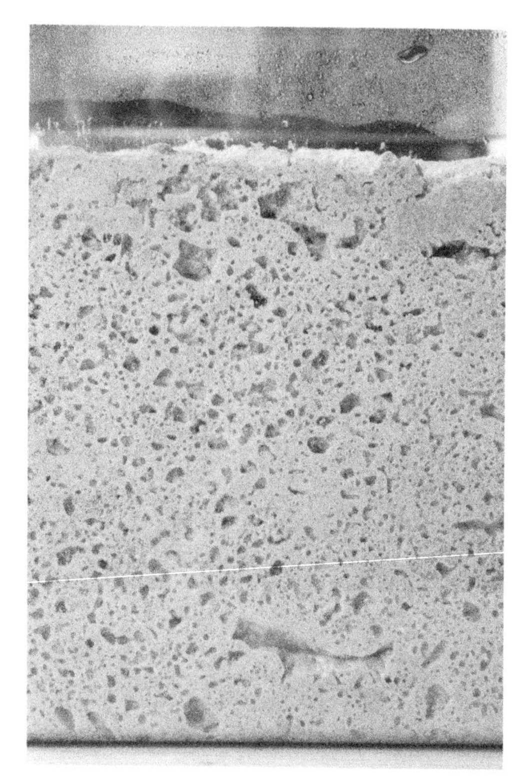

Rye Refresher Sponge will be cracked on the top, and very bubbly under the surface.

Because this dough is made from 100% rye flour, it will not develop gluten. Mix the dough until it is homogenous.

Wet the dough on a damp towel before rolling in oats.

The finished bread will be very dense. Do not freeze this bread; it will hold under refrigeration for several weeks.

Kartoffelbrot

DESIRED DOUGH TEMPERATURE: 75°F/24°C

INGREDIENT	US	METRIC	BAKERS' PERCENTAGE
Water	5 lb 8 oz	2.5 kg	50.0%
Bread flour	11 lb	5 kg	100.0%
Mashed potatoes	1 lb 10½ oz	750 g	15.0%
Sea salt	3½ oz	100 g	2.0%
Sugar	1¾ oz	50 g	1.0%
Yeast, instant dry	6 oz	175 g	3.5%
Butter, soft	2½ oz	75 g	1.5%
Rye Sourdough (page 147)	1 lb 10½ oz	750 g	15.0%
Total Dough Weight	20 lb 12 oz	9.4 kg	188.0%

1. Combine the water, flour, potatoes, salt, sugar, yeast, butter, and rye sourdough in a mixer. Mix on low speed until homogenous, about 3 minutes. Increase the speed and mix on high until the dough reaches the improved stage of gluten development, about 2 minutes.
2. Remove the dough from the mixer and bulk-ferment, covered, until doubled in size, about 1½ hours. Fold the dough halfway through fermentation.
3. Divide the dough into 1 lb 10½ oz/750 g pieces and preshape into rounds (see Note). Bench-rest the dough for 20 minutes.
4. Shape the dough into 10-in/2.5-cm bâtards and place in lightly floured bannetons. Proof the dough for 1 hour or as needed, depending on the environment.
5. Score the dough and load into a 470°F/243°C oven. Steam for 1 second before loading and 3 seconds after loading. Lower the temperature to 460°F/238°C and bake for 30 minutes. Open the vent and bake for an additional 20 minutes, until dark golden brown.
6. Place the finished breads on a rack to cool.

CHEF'S NOTE

Cheese and/or herbs can be added to this dough. If desired, add 0.5% cheese and 0.5 % herbs.

Rheinisches Schwarzbrot

DESIRED DOUGH TEMPERATURE: 82°F/28°C

INGREDIENT	US	METRIC	BAKERS' PERCENTAGE
Refresher Sponge			
Water	8½ oz	240 g	1.7%
Rye Sourdough (page 147)	3½ oz	100 g	0.7%
Coarse rye flour	7 oz	200 g	1.4%
Base Levain			
Water	2 lb 2 oz	960 g	6.8%
Refresher Sponge	1 lb 3 oz	540 g	3.8%
Coarse rye flour	3 lb 8½ oz	1.6 kg	11.4%
Rye Berry Soaker			
Water	8 lb 13 oz	4 kg	28.6%
Whole rye berries, broken	8 lb 13 oz	4 kg	28.6%
Full Levain			
Water	9 lb 11¼ oz	4.4 kg	31.4%
Base Levain	6 lb 13½ oz	3.1 kg	22.1%
Medium coarse rye flour	11 lb 7¾ oz	5.2 kg	37.1%
Final Dough			
Water	3 lb 8½ oz, or as needed	1.6 kg, or as needed	11.4%
Full Levain (see Note)	27 lb 12½ oz	12.6 kg	90.0%
Rye Berry Soaker	17 lb 10 oz	8 kg	57.1%
Medium coarse rye flour	15 lb 7 oz	7 kg	50.0%
Salt	10½ oz	300 g	2.1%
Total Dough Weight	65 lb 2 oz	29.5 kg	210.6%
Rolled oats	As needed	As needed	

1. Prepare the refresher sponge 2 days before baking the final dough. Combine the water, rye sourdough, and flour and mix until homogenous. The desired dough temperature is 75°F/24°C. Store covered at room temperature for 7 hours.

2. Prepare the base levain. The desired dough temperature is 77°F/25°C. Add the water, refresher sponge, and flour to a mixer on low speed for 2 minutes, until homogenous. Cover and store at room temperature for 10 to 12 hours.

3. Prepare the rye berry soaker a day before preparing the final dough. Heat the water to 85°F/29°C and pour over the rye berries. Cover and allow to soak overnight.

4. The following day, prepare the full levain. Add the water, base levain, and flour to a mixer, and mix on low speed for 10 minutes, until homogenous. Cover and allow to sit at room temperature for 4 hours.

5. To mix the final dough, combine the water, full levain, rye berry soaker, rye flour, and salt in a mixer. Mix on low speed until the dough comes together and is homogenous, about 8 minutes.

6. Remove the dough from the bowl and place on a floured surface. Cover with plastic and bench-rest the dough. The dough is ready to divide when it begins to pucker, after approximately 20 to 30 minutes.

7. Divide the dough into 4 lb 6½ oz/2.0 kg pieces and preshape into rounds. Place the rounds seam side up on the surface, cover, and rest for 10 to 15 minutes. Meanwhile, lightly oil Pullman loaf pans.

8. Spread the oats on a half sheet tray and line another half sheet tray with a wet towel. Shape each piece of dough into a log, roll on the wet towel, and then in the oats. Place the loaves seam side down in the prepared pans. Proof for 2 to 3 hours or as needed, depending on the environment.

9. Lightly score the loaves once along the length of the bread. Load into a 480°F oven and steam for 1 second before and after loading.

10. Bake at 480°F/249°C for approximately 10 minutes, opening the dampers approximately 5 minutes into baking. Drop the temperature to 400°F/204°C for the remainder of baking, approximately 2 hours.

11. Remove the breads from the pans and place the finished breads on a rack to cool.

CHEF'S NOTE

Save 3½ oz/100 g of Full Levain to use in the refresher sponge for the next baking.

To make "real" Pumpernickel, bake the dough for 24 to 36 hours at 210°F/99°C.

This bread is best if allowed to age for at least a day before slicing.

Landbrot

DESIRED DOUGH TEMPERATURE:
84 TO 86°F/29 TO 30°C

INGREDIENT	US	METRIC	BAKERS' PERCENTAGE
Refresher Sponge			
Water	3½ oz	100 g	2.0%
Rye Sourdough (page 147)	3½ oz	100 g	2.0%
Fine dark rye flour	3 oz	85 g	1.7%
Base Levain			
Water	10½ oz	300 g	6.0%
Refresher Sponge	10 oz	285 g	5.7%
Fine dark rye flour	1 lb 4½ oz	580 g	11.6%
Full Levain			
Water	3 lb 1½ oz	1.4 kg	28.1%
Base Levain	2 lb 3¼ oz	1.1 kg	23.4%
Fine dark rye flour	2 lb 14 oz	1.3 kg	26.6%
Final Dough			
Water	4 lb 3 oz	1.9 kg	37.0%
Full Levain	8 lb 6 oz	3.8 kg	76.0%
Bread flour	3 lb 5 oz	1.5 kg	30.0%
Fine dark rye flour	3 lb 5 oz	1.5 kg	30.0%
Salt	4¼ oz	120 g	2.4%
Total Dough Weight	19 lb 7¼ oz	8.8 kg	175.4%

1. Prepare the refresher sponge at least a day before baking the final dough. For the refresher sponge, combine the water, rye sourdough, and flour and mix until homogenous. Store, covered, at room temperature for 6 hours. The desired dough temperature is 75°F/24°C.

2. For the base levain, combine the water, refresher sponge, and flour and mix until homogenous. Store, covered, at room temperature for 8 to 16 hours. The desired dough temperature is 75°F/24°C.

3. The following day, prepare the full levain. Combine the water, base levain, and flour, and mix until homogenous. Store, covered, at room temperature for 3 hours. The desired dough temperature is 84 to 86°F/29 to 30°C.

4. For the final dough, combine the water, full levain, flours, and salt in a mixer, and mix on low speed for 8 minutes until a dough is formed.

5. Remove the dough from the mixer and place on a floured board. Store, covered, at room temperature for 20 to 30 minutes, until puckering occurs.
6. Divide the dough into 1 lb 10 oz/750 g pieces and preshape into rounds. Bench-rest the dough for 10 to 15 minutes.
7. Shape the dough into 10-in/25-cm bâtards and place seam side up in lightly floured bannetons. Proof for 1 to 2 hours or as needed, depending on the environment.
8. Score and load the breads into oven, steaming 1 second before loading and 3 seconds after loading.
9. Bake at 480°F/249°C for approximately 10 minutes, opening the damper after about 5 minutes. Lower the temperature to 400°F/204°C and bake for 45 minutes.
10. Place the finished breads on a rack to cool.

Haferbrötchen

DESIRED DOUGH TEMPERATURE: 78°F/26°C

INGREDIENT	US	METRIC	BAKERS' PERCENTAGE
Oat Flake Soaker			
Water, cold	10½ oz	300 g	100.0%
Rolled oats, lightly toasted	10½ oz	300 g	100.0%
Final Dough			
Water	3 lb 5 oz	1.5 kg	59.0%
Bread flour	4 lb	1.8 kg	70.0%
Yeast, instant dry	4½ oz	125 g	5.0%
Whole milk powder	2¾ oz	75 g	3.0%
Diastatic dry malt	2¾ oz	75 g	3.0%
Lard	2¾ oz	75 g	3.0%
Oat Flake Soaker	1 lb 5 oz	600 g	24.0%
Rolled oats, lightly toasted	1 lb 10½ oz	750 g	30.0%
Sea salt	1¾ oz	50 g	2.0%
Total Dough Weight	10 lb 15 oz	5 kg	199.0%
Rolled oats	As needed	As needed	

1. Prepare the soaker. Combine the water and oats in a bowl and allow to sit at room temperature for 1 hour.
2. To prepare the final dough, combine the water, flour, yeast, milk powder, malt, lard, and soaker in a mixer. Mix on low speed until homogenous, about 3 minutes. Increase the speed to high and mix until the dough reaches the improved stage of gluten development, about 1 minute.
3. Add the oats and salt, and mix on low speed until incorporated, about 1 minute. Remove the dough from the mixer and bulk-ferment, covered, for about 45 minutes.
4. Divide the dough into 3½ oz/100 g pieces and preshape into rounds. Bench-rest the dough for 10 to 15 minutes.
5. Shape the dough into bâtards and moisten on a damp towel-lined sheet tray. Roll the bâtards in the oats and place seamside down into lightly oiled strap pans. Proof for about 1 hour or as needed, depending on the environment.

6. Score the dough and load into a 440°F/227°C oven. Steam, and bake until golden brown, about 45 minutes.

7. Remove the breads from the pans immediately after baking. Place the finished breads on a rack to cool.

The soaker can be mixed by hand using cold water.

Haferbrötchen dough before final fermentation.

Haferbrötchen after final fermentation.

The finished bread will be lighter in color than some of the other rye breads, but will still have a tight, dense interior.

Landbrot mit Sauerkraut

DESIRED DOUGH TEMPERATURE: 82°F/28°F

INGREDIENT	US	METRIC	BAKERS' PERCENTAGE
Water	5 lb 12 oz	2.6 kg	58.0%
Bread flour	7 lb 11 oz	3.5 kg	78.0%
Whole wheat flour	1 lb 1½ oz	500 g	11.0%
Medium course rye flour	1 lb 1½ oz	500 g	11.0%
Sea salt	4½ oz	125 g	2.8%
Yeast, instant dry	3¾ oz	105 g	2.3%
Rye Sour (page 147)	2 lb 3¼ oz	1 kg	22.0%
Sauerkraut, drained	4 lb 3 oz	1.9 kg	42.0%
Total Dough Weight	22 lb 10 oz	10.3 kg	227.1%

1. Combine the water, bread flour, whole wheat flour, rye flour, sea salt, yeast, and rye sour in a mixer bowl. Mix on low speed until a dough forms, about 6 minutes. Add the sauerkraut and mix for an additional 2 minutes.
2. Transfer the dough to a floured board and cover. Bench-rest until the dough begins to pucker, 20 to 30 minutes.
3. Divide the dough into 2 lb 3¼ oz/1 kg pieces and preshape into rounds. Bench-rest the dough for 10 to 15 minutes, with the seams facing up.
4. Shape the dough into 10-in/25-cm bâtards and place seam side up in lightly floured bannetons. Proof for 1 to 2 hours or as needed, depending on the environment.
5. Score the dough. Load the loaves into a 500°F/260°C hearth oven, and steam for 1 second before loading and 3 seconds after loading. Immediately lower the temperature to 480°F/249°C and bake for 10 minutes, opening the damper about 5 minutes into the baking. Lower the temperature to 400°F/204°C and bake for an additional 50 minutes, until golden brown.
6. Place the finished breads on a rack to cool.

Leinsamenbrot

DESIRED DOUGH TEMPERATURE: 80°F/27°C

INGREDIENT	US	METRIC	BAKERS' PERCENTAGE
Soaker			
Water	6 lb 10 oz	3 kg	30.0%
Flaxseed	2 lb 3¼ oz	1 kg	10.0%
Final Dough			
Water	2 lb 14 oz	1.3 kg	13.0%
Medium rye flour	4 lb 6½ oz	2 kg	20.0%
Bread flour	8 lb 13 oz	4 kg	40.0%
Salt	6¼ oz	175 g	1.8%
Yeast, instant dry	5¼ oz	150 g	1.5%
Rye Sourdough (page 147)	15 lb 14¼ oz	7.2 kg	72.0%
Soaker	8 lb 13 oz	4 kg	40.0%
Total Dough Weight	41 lb 9 oz	18.8 kg	188.3%

1. Prepare the soaker. Combine the flaxseed with the water in a bowl and let sit for 1 hour.
2. Make the final dough. Combine the water, rye flour, bread flour, salt, yeast, and rye sourdough in a mixer. Mix on low speed for 8 minutes. As the dough comes together, add the soaker. The dough should be very well hydrated.
3. Transfer the dough to a floured board. Cover and bench-rest until the dough begins to pucker, 20 to 30 minutes.
4. Divide the dough into 1 lb 10½ oz/750 g pieces and preshape into rounds. Bench-rest the dough for 10 to 15 minutes, with the seams facing up.
5. Shape the rounds into boules or bâtards, and place them seam side up into lightly floured bannetons. Proof for 1 to 2 hours or as needed, depending on the environment.
6. Score the dough and load into a 480°F/249°C oven. Steam 1 second before loading and 5 seconds after loading. Bake for 5 minutes, before opening the vent and baking for an additional 5 minutes. Lower the temperature to 400°F/204°C and bake until golden brown, about 35 minutes.
7. Place the finished breads on a rack to cool.

Rye and Onion Sourdough

DESIRED DOUGH TEMPERATURE: 80°F/27°C

INGREDIENT	US	METRIC	BAKERS' PERCENTAGE
Onion Rye Sour			
Water	1 lb 12¾ oz	815 g	16.3%
Medium rye flour	2 lb 10 oz	1.2 kg	24.1%
Salt	1 oz	30 g	0.5%
Onions, medium diced	1 lb 7½ oz	670 g	13.4%
Rye Sourdough (page 147)	6 oz	170 g	3.5%
Final Dough			
Water	4 lb 13½ oz	2.2 kg	44.1%
Medium rye flour	4 lb	1.8 kg	35.8%
Bread flour	4 lb 6½ oz	2 kg	40.1%
Yeast, instant dry	2½ oz	75 g	1.5%
Salt	3¾ oz	105 g	2.1%
Malt syrup	1 oz	30 g	0.5%
Onions, medium diced, sautéed, and cooled	1 lb 7½ oz	670 g	13.4%
Onion Rye Sour	6 lb 6 oz	2.9 kg	57.2%
Total Dough Weight	21 lb 7½ oz	9.7 kg	194.8%

1. Combine all of the ingredients for the Onion Rye Sour and allow to ferment for 16 hours. To make the final dough, combine the water, rye flour, bread flour, yeast, salt, malt syrup, onions, and onion rye sour in a mixer and mix on low speed until homogenous, about 4 minutes. Increase the speed to high and mix until the dough has reached the improved stage of gluten development, about 3 minutes.

2. Remove the dough from the mixer and bulk-ferment, covered, until doubled in size, about 30 minutes.

3. Divide the dough into 1 lb 4 oz/570 g pieces and preshape into rounds. Immediately shape into 10-in/25-cm bâtards and place in lined and floured bannetons. Proof the dough for 30 to 40 minutes or as needed, depending on the environment.

4. Load the dough into a 490°F/254°C oven and immediately lower the temperature to 480°F/249°C. Bake until deep golden brown, about 18 minutes, followed by an additional 12 minutes with the vent open.

5. Place the finished breads on a rack to cool.

Cheddar Onion Rye Rolls

DESIRED DOUGH TEMPERATURE: 77°F/25°C

INGREDIENT	US	METRIC	BAKERS' PERCENTAGE
Water	6 lb 6 oz	2.9 kg	58.6%
Rye Sourdough (page 147)	1 lb 4-¼ oz	575 g	11.5%
Malt syrup	1 oz	30 g	0.5%
Vegetable oil	3¼ oz	90 g	1.8%
Molasses	3¼ oz	90 g	1.8%
Medium rye flour	2 lb 10 oz	1.2 kg	23.8%
Bread flour	8 lb 6 oz	3.8 kg	76.2%
Yeast, instant dry	1 oz	30 g	0.6%
Sugar	3¼ oz	90 g	1.8%
Salt	4½ oz	130 g	2.6%
Cheddar, grated	2 lb 3¼ oz	1 kg	20.0%
Onions, medium diced, sautéed	2 lb 3¼ oz	1 kg	20.0%
Total Dough Weight	24 lb 2½ oz	12 kg	219.2%
Grated cheddar cheese, for garnish	2lb 3¼ oz	1 kg	20.0%
Salt	As needed	As needed	

1. Combine the water, rye sourdough, malt syrup, vegetable oil, molasses, rye flour, bread flour, yeast, sugar, and salt in a mixer. Mix on low speed until homogenous, about 4 minutes. Increase the speed to high and mix until the dough reaches the improved level of gluten development, about 4 minutes.
2. Add the cheese and onions, and mix on low speed until incorporated, about 2 minutes.
3. Remove the dough from the mixer and bulk-ferment, covered, until doubled in size, about 45 minutes.
4. Divide the dough into 4 lb/1.8 kg pieces and preshape into round presses. If using a dough divider, this will give you 36 pieces at 1¾ oz/50 g each. Bench-rest the dough for 15 minutes.

5. Dust the presses with flour, and divide the presses on a dough divider. Use the machine to round the rolls. Place the rolls on a parchment-lined sheet tray. Stagger the rolls to fit 6 rows of 5 each. Proof the rolls for 35 to 40 minutes or as needed, depending on the environment.
6. To finish, score each roll and spritz with water. Top with cheese and a light sprinkle of salt.
7. Load the rolls into a 475°F/246°C convection oven, and immediately lower the temperature to 425°F/218°C. Steam with 6 cups/1.5 L water and bake for 14 minutes, until the rolls are golden brown. Open the vents and bake for an additional 2 minutes.
8. Cool the rolls on the sheet trays.

Vinchgauer

DESIRED DOUGH TEMPERATURE: 85°F/29°C

INGREDIENT	US	METRIC	BAKERS' PERCENTAGE
Water	4 lb 6½ oz	2 kg	83.3%
Medium coarse rye flour	3 lb 1½ oz	1.4 kg	58.3%
Bread flour	2 lb 3¼ oz	1 kg	41.7%
Rye Sourdough (page 147)	3 lb 8½ oz	1.6 kg	66.6%
Yeast, instant dry	1¾ oz	50 g	2.1%
Salt	2½ oz	75 g	3.1%
Coriander seeds, toasted and ground	½ oz	10 g	0.4%
Fennel seeds, toasted	½ oz	10 g	0.4%
Caraway seeds, toasted	¾ oz	20 g	0.8%
Total Dough Weight	22 lb 10 oz	10.3 kg	256.7%
Rye flour	As needed	As needed	

1. Combine the water, rye flour, bread flour, rye sourdough, yeast, salt, and coriander, fennel, and caraway in a mixer, and mix on low speed until a homogenous dough forms, about 8 minutes.
2. Remove the dough from the mixer and place on a floured board. Bulk-ferment, covered, until the dough begins to pucker, 20 to 30 minutes.
3. Divide the dough into 7 oz/200 g pieces, and preshape into rounds. Bench-rest the dough for 10 to 15 minutes, with the seam side up.
4. Gently flatten the rounds into discs, roll in some rye flour, and place seam side up into couches. Proof for about 1 hour or as needed, depending on the environment.
5. While the dough is proofing, soak about 6 oz/170 g wood chips in water. Place the wood chips near the back of the oven about 5 minutes before baking.
6. Load the dough into a 480°F/249°C oven. Steam for 1 second before loading and 3 seconds after loading. Bake for about 10 minutes, opening the vent halfway through. Lower the temperature to 400°F/204°C and bake until dark brown, about 45 minutes (see Note).
7. Place the finished breads on a rack to cool.

CHEF'S NOTE

Make sure to sweep the oven after baking this bread to remove wood chips and ash.

Whole-Grain Rye Bread

DESIRED DOUGH TEMPERATURE: 80°F/27°C

INGREDIENT	US	METRIC	BAKERS' PERCENTAGE
Cracked Rye Preferment			
Water, 90°F/32°C	1 lb 10 oz	735 g	28.3%
Rye Sourdough (page 147)	6¼ oz	175 g	6.8%
Cracked rye berries	1 lb 12 oz	800 g	30.8%
Cracked Rye Soaker			
Water, 90°F/32°C	5 oz	140 g	5.4%
Salt	2 oz	55 g	2.1%
Cracked rye berries	14 oz	390 g	15.0%
Rye Berry Soaker			
Water	1 lb 8 oz	680 g	26.1%
Rye berries	9¼ oz	260 g	10.1%
Final Dough			
Water, 100°F/38°C	1 lb	450 g	17.4%
Cracked rye berries	2 lb 1½ oz	950 g	36.5%
Sunflower seeds	5½ oz	155 g	7.0%
Medium rye flour	7 oz	200 g	7.7%
Yeast, instant dry	¼ oz	10 g	0.3%
Total Dough Weight	11 lb	5 kg	193.5%
Vegetable oil	As needed	As needed	

1. Prepare the cracked rye preferment, cracked rye soaker, and rye berry soaker the day before preparing the final dough.
2. To prepare the cracked rye preferment, combine the water, rye sourdough, and cracked rye berries in a mixer and mix on low speed for 15 minutes. Place in a covered container and reserve.
3. To make the cracked rye soaker, mix the water, salt, and cracked rye berries by hand in a bowl until combined. Place in a covered container and reserve.
4. To make the rye berry soaker, bring the water to a boil in a medium saucepan. Add the whole rye berries and boil, covered, for 10 minutes. Remove from the heat and soak the berries overnight. Place in a covered container and reserve.
5. To prepare the final dough, combine the cracked rye preferment, cracked rye soaker, and rye berry soaker with the water, cracked rye berries, sunflower seeds, flour, and yeast in a mixer. Mix with the paddle attachment for 15 minutes. Every 5 minutes, check

the dough's hydration and add water if it begins to seem dry. The dough should be wet, but not sticky.

6. Remove the dough from the mixer and bulk-ferment, covered, for 10 minutes.

7. Divide the dough into 1 lb 8 oz/680 g pieces and shape into rounds. Place 2 rounds in each loaf pan, brushing the loaves with oil to keep them separated. Proof for 50 to 60 minutes or as needed, depending on the environment.

8. Load the loaves into a 470°F/243°C oven and immediately lower the temperature to 440°F/227°C. Bake until the dough reaches an internal temperature of 205°F/96°C, for 45 to 60 minutes.

9. Remove the finished breads from the pans and place on a rack to cool.

Rye Multigrain Bread

DESIRED DOUGH TEMPERATURE: 78°F/26°C

INGREDIENT	US	METRIC	BAKERS' PERCENTAGE
Soaker			
Water, room temperature	6 lb 10 oz	3.0 kg	40.9%
9-grain mix	2 lb 7 oz	1.1 kg	15.3%
Flaxseed	1 lb 15 oz	855 g	11.5%
Sunflower seeds	14½ oz	415 g	5.6%
Final Dough			
Water	4 lb 1 oz	1.8 kg	24.8%
Bread flour	6 lb 13 oz	3.1 kg	41.5%
Medium rye flour	4 lb 4 oz	1.9 kg	26.0%
Salt	6 oz	170 g	2.3%
Yeast, instant dry	2 oz	55 g	0.7%
Rye Sourdough (page 147)	5 lb	2.3 kg	30.6%
Malt syrup	1½ oz	40 g	0.5%
Grain Soaker	11 lb 11 oz	5.3 kg	71.7%
Total Dough Weight	32 lb 5½ oz	14.7 kg	198.1%
Sesame seeds or rolled oats	As needed	As needed	

1. Prepare the soaker 18 hours before mixing the final dough. Combine the water, 9-grain mix, flaxseed, and sunflower seeds in a bowl and mix until blended. Place in a covered container and store at room temperature for 18 hours.

2. To prepare the final dough, combine the water, bread flour, rye flour, salt, yeast, rye sourdough, and malt syrup in a mixer. Mix on low speed until homogenous, about 3 minutes. Increase the speed to high and mix for about 2 minutes.

3. Add half the soaker and mix on low speed to incorporate, about 2 minutes. Add the remaining soaker and mix until homogenous, about 2 minutes. Increase the speed to high and mix until the dough reaches the improved stage of gluten development, about 2 minutes.

4. Remove the dough from the mixer and bulk-ferment, covered, for about 30 minutes.

5. Divide the dough into 1 lb 4 oz/570 g pieces and preshape into rounds. Bench-rest the dough for 10 to 15 minutes.

6. Shape into 10-in/25-cm bâtards (see Note). Roll in sesame seeds or oats and place in lined bannetons. Proof the dough for about 30 minutes or as needed, depending on the environment.

7. Load into a 485°F/252°C oven and immediately lower the temperature to 475°F/246°C. Bake until lightly brown, about 18 minutes. Open the vent and bake until golden brown, about 12 minutes.

8. Place the finished breads on a rack to cool.

CHEF'S NOTE

Ulmer Spatz is a shape that can be used with many varieties of dough, such as for pretzels or whole wheat loaves. To shape instead into Ulmer Spatz, roll the preshaped dough into a 20-in/50-cm log. Fold the dough over and make a knot at one end of the log. Proof the dough for 30 minutes or as needed, depending on the environment. Load into a 480°F/249°C oven and immediately lower the temperature to 460°F/238°C. Bake until golden brown, about 20 minutes. Open the vent and bake until golden brown, about 12 minutes.

St. Gallener's Brot

DESIRED DOUGH TEMPERATURE: 84°F/29°F

INGREDIENT	US	METRIC	BAKERS' PERCENTAGE
Water	4 lb 13½ oz	2.2 kg	62.9%
Rye Sourdough (page 147)	6 lb	2.7 kg	77.1%
Bread flour	5 lb 8 oz	2.5 kg	71.4%
Fine white rye flour	2 lb 3¼ oz	1.0 kg	28.6%
Salt	4¼ oz	120 g	3.4%
Total Dough Weight	18 lb 11 oz	8.5 kg	243.4%

1. Combine the water, rye sourdough, bread flour, rye flour, and salt in a mixer and mix on low speed until homogenous, about 4 minutes. Increase the speed and mix on high speed until the dough reaches the improved stage of gluten development, about 2 minutes.
2. Remove the dough from the mixer and bulk-ferment, covered, until doubled in size, about 1 hour.
3. Scale into 1 lb 8 oz/680 g pieces and preshape into oblongs. Bench-rest the dough for about 10 minutes.
4. Preshape the dough into a 16-in/40-cm loaf. Place in a floured couche. Proof the dough for about 1 hour or as needed, depending on the environment.
5. Transfer the dough to a loader and shape into a coil, tucking the last quarter of the loaf underneath the rest of the coil, like a snail.
6. Load the dough into a 470°F/243°C oven, and immediately drop the temperature to 440°F/227°C. Bake until the bread is golden brown, 45 to 60 minutes.
7. Place the finished loaf on a rack to cool.

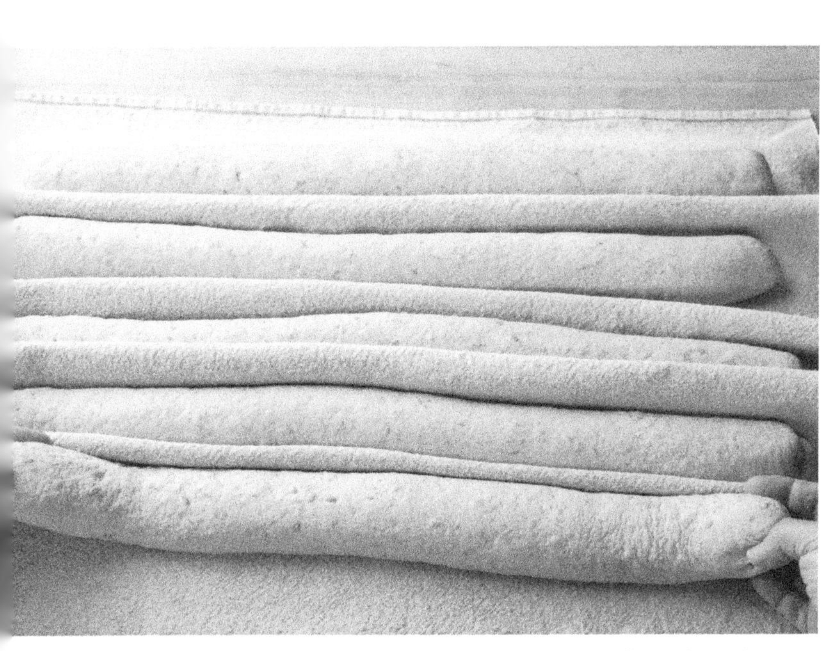

Shape the dough into long loaves and place on a floured couche.

Shape each loaf into a coil.

Tuck the end of the dough under the coil, so that the top is resting on the tail end.

The finished bread is brushed with a flour-and-water mixture that gives the bread its signature dark color and shine.

Caraway Rye Bread

DESIRED DOUGH TEMPERATURE: 77°F/25°C

INGREDIENT	US	METRIC	BAKERS' PERCENTAGE
Water	7 lb 4½ oz	3.3 kg	65.0%
Bread flour	9 lb 4 oz	4.2 kg	84.8%
White rye flour	1 lb 11 oz	765 g	15.4%
Rye Sourdough (page 147)	1 lb 8 oz	680 g	13.5%
Yeast, instant dry	1¾ oz	50 g	1.0%
Sea salt	3¼ oz	90 g	1.8%
Caraway seeds	1½ oz	40 g	0.8%
Total Dough Weight	20 lb 1½ oz	9.1 kg	182.3%

1. Combine the water, bread flour, rye flour, and rye sourdough in a mixer. Mix on low speed until homogenous, about 6 minutes. Increase the speed to high and mix for 30 seconds. Reduce the speed to low and mix until the dough is smooth, about 1½ minutes.
2. Stop the mixer and sprinkle the yeast on top of the dough. Rest the dough in the mixer to autolyse for about 15 minutes.
3. Return the mixer to low speed to incorporate the yeast. Mix for 1½ minutes, add the salt, and mix for an additional 30 seconds. Increase the speed and mix on high until the dough has reached the improved stage of gluten development, about 2 minutes.
4. Return the mixer to low speed, add the caraway seeds, and mix until the seeds are incorporated, about 1 minute.
5. Remove the dough from the mixer and bulk-ferment, covered, for 1½ hours. Fold the dough halfway through fermentation.
6. Divide the dough into 1 lb 8¾ oz/700 g pieces and preshape into rounds. Bench-rest the dough for 10 to 15 minutes.
7. Shape the dough into 10-in/2.5-cm bâtards or boules. Place the dough seam side up in floured bannetons. Proof for 1 hour or as needed, depending on the environment.
8. Score the loaves. Load the bread into a 450°F/232°C hearth oven, and immediately lower the temperature to 440°F/227°C. Bake until the bread is golden brown, about 18 minutes, followed by an additional 15 minutes with the vent open.
9. Place the finished breads on a rack to cool.

Rye and Sunflower Bread

DESIRED DOUGH TEMPERATURE: 76°F/24°C

INGREDIENT	US	METRIC	BAKERS' PERCENTAGE
Water	7 lb 15 oz	3.6 kg	72.5%
Medium rye flour	6 lb 2½ oz	2.8 kg	56.3%
Bread flour	4 lb 13½ oz	2.2 kg	43.8%
Yeast, instant dry	1 oz	30 g	0.6%
Rye Sourdough (page 147)	4 lb 10 oz	2.1 kg	41.9%
Malt syrup	1 oz	30 g	0.5%
Sunflower seeds, toasted	2 lb 14 oz	1.3 kg	25.0%
Salt	4½ oz	125 g	2.5%
Total Dough Weight	26 lb 12½ oz	12.2 kg	243.1%

1. Combine the water, rye flour, bread flour, yeast, rye sour, malt syrup, sunflower seeds, and salt in a mixer. Mix on low speed until homogenous, 8 to 12 minutes until the dough reaches the improved stage of gluten development.
2. Remove the dough from the mixer and bulk-ferment, covered, until doubled in size, about 30 minutes.
3. Scale the dough into 1 lb 4 oz/570 g pieces and preshape into rounds. Bench-rest the dough for 10 to 15 minutes.
4. Shape the dough into 10-in/25-cm bâtards. Place in lightly floured baskets. Proof for 30 to 35 minutes or as needed, depending on the environment.
5. Bake in a 465°F/241°C oven until golden brown, 40 to 45 minutes.
6. Place the finished breads on a rack to cool.

Basler Brot

DESIRED DOUGH TEMPERATURE: 82°F/28°C

INGREDIENT	US	METRIC	BAKERS' PERCENTAGE
Water	4 lb 6½ oz	2.0 kg	66.7%
Rye Sourdough (page 147)	4 lb 6½ oz	2.0 kg	66.7%
Bread flour	5 lb 8 oz	2.5 kg	83.3%
Medium fine rye flour	1 lb 2 oz	510 g	16.7%
Salt	4¼ oz	120 g	4.0%
Yeast, instant dry	1¾ oz	50 g	1.7%
Total Dough Weight	15 lb 14 oz	7.2 kg	239.1%

1. Combine the water, rye sourdough, bread flour, rye flour, salt, and yeast in a mixer. Mix on low speed until homogenous, about 4 minutes. Increase the speed to high and mix until the dough reaches the improved stage of gluten development, about 3 minutes.
2. Remove the dough from the mixer and bulk-ferment, covered, until doubled in size, about 45 minutes.
3. Divide the dough into 10 oz/285 g pieces and pre-shape into rounds. Bench-rest the dough for 15 to 20 minutes.
4. Shape the dough into rounds and place 4 to 5 in a row on floured couches. The rounds should touch. Proof for about 1 hour or as needed, depending on the environment.
5. Transfer the dough to a loader as one piece, and score across the entire row.
6. Load the dough into a 490°F/254°C oven and immediately lower the temperature to 480°F/249°C. Bake until deep golden brown, about 18 minutes, followed by an additional 12 minutes with the vent open.
7. Place the finished bread on a rack to cool.

Basler Brot dough should be very elastic and not overly hydrated.

Preshape the dough into rounds.

Lay the rounds in tight rows of 5 on a couche.

Flour the dough before scoring.

Score the rows in one long motion, rather than as individual loaves.

The adjoined pieces are sold as a whole, to be divided at home.

Krauterquarkbrot

DESIRED DOUGH TEMPERATURE: 77°F/25°C

INGREDIENT	US	METRIC	BAKERS' PERCENTAGE
Water	4 lb 13½ oz	2.2 kg	54.6%
Rye Sour	4 lb 3 oz	1.9 kg	47.2%
Bread flour	7 lb 15 oz	3.6 kg	88.7%
Rye flour	1 lb	450 g	11.3%
Yeast, instant dry	4½ oz	125 g	3.1%
Sea salt	3¼ oz	90 g	2.2%
Quark	1 lb 1½ oz	500 g	12.6%
Herbs (mixture of parsley, thyme, rosemary, and chervil. minced)	2½ oz	70 g	1.8%
Total Dough Weight	19 lb 8 oz	8.8 kg	221.5%

1. Combine the water, rye sour, bread flour, rye flour, yeast, and salt in a mixer. Mix on low speed for 7 minutes. Add the quark and the herbs and mix until the dough has formed, about 1 minute
2. Transfer the dough to a floured board. Cover and bulk-ferment for 20 to 30 minutes, until the dough puckers.
3. Divide the dough into 1 lb 7 oz/650 g pieces and preshape into rounds (see Note). Bench-rest the dough for 10 to 15 minutes with the seam facing up.
4. Shape the dough into bâtards and place seam side up in lightly floured bannetons. Proof for about 1 hour or as needed, depending on the environment.
5. Score the bread and load into a 480°F/249°C oven. Steam for 1 second before loading and 3 seconds after loading. Bake for 5 minutes, open the vent, and bake for an additional 5 minutes. Lower the temperature to 400°F/204°C and bake until dark golden brown, an additional 50 minutes.
6. Place the finished breads on a rack to cool.

CHEF'S NOTE

The dough can also be divided into 3 oz/85 g pieces for rolls. Proceed with Step 5 of the recipe but adjust the baking time.

Three Step Sour

STEP 1: REFRESHER SPONGE

FERMENTATION TIME: 7 HOURS

DESIRED DOUGH TEMPERATURE: 75°F/24°C

INGREDIENT	US	METRIC	BAKERS' PERCENTAGE
Water	8½ oz	240 g	1.7%
Rye Sourdough (page 147)	3½ oz	100 g	0.7%
Medium coarse rye flour	7 oz	200 g	1.4%
Total	**1 lb 3 oz.**	**540 g**	**3.8%**

Method:

1. Combine the water, rye sourdough, and flour and mix until homogenous.
2. Store covered at room temperature for 7 hours.

STEP 2: BASE LEVAIN

FERMENTATION TIME: 10–12 HOURS

DESIRED DOUGH TEMPERATURE: 77°F/25°C

INGREDIENT	US	METRIC	BAKERS' PERCENTAGE
Water	2 lb 2 oz	960 g	6.8%
Refresher Sponge	1 lb 3 oz	540 g	3.8%
Medium coarse rye flour	3 lb 8½ oz	1.6 kg	11.4%
Total	**6 lb 13½ oz.**	**3.1 kg**	**22%**

Method:

1. Add the water, refresher sponge, and flour to a mixer, and mix on low speed for 2 minutes, until homogenous.
2. Store covered at room temperature for 10–12 hours.

STEP 3: FULL LEVAIN

FERMENTATION TIME: 4 HOURS

DESIRED DOUGH TEMPERATURE: 84–86°F/28–30°C

INGREDIENT	US	METRIC	BAKERS' PERCENTAGE
Water	9 lb 11¼ oz	4.4 kg	31.4%
Base Levain	6 lb 13½ oz	3.1 kg	22.1%
Medium coarse rye flour	11 lb 7¾ oz	5.2 kg	37.1%
Total	**28 lb ½ oz**	**12.7 kg**	**90.6%**

Method:

1. Add the water, base levain, and flour to a mixer, and mix on low speed for 10 minutes, until homogenous.
2. Cover and allow to sit at room temperature for 4 hours.

CHEF'S NOTE

After using Full Levain, save any excess to use in the refresher sponge for future baking.

Decorative dough, or dead dough, is an unleavened mixture of rye flour, buckwheat flour, and simple syrup. It can be used as a decorative addition to breads in a display case or bakery window. The dough, sometimes colored with natural ingredients such as chili powder and turmeric, can be shaped or embellished in any number of ways, and it dries hard with a glossy exterior. Often used in competitive showpiece designs, decorative dough is an excellent way to catch a customer's eye and draw attention to an otherwise ordinary loaf of bread. A visually appealing showpiece might also attract customers to your bakeshop.

Syrup for Decorative Dough

INGREDIENT	US	METRIC	BAKERS' PERCENTAGE
Water, boiling	9 lb 6½ oz	4.2 kg	100.0%
Sugar	9 lb 11 oz	4.4 kg	103.2%
Glucose syrup	3 lb 8 oz	1.6 kg	37.2%
Yield	22 lb 9½ oz	10.3 kg	240.4%

Add the boiling water to the sugar and glucose. Stir until the sugar is dissolved. Cover and store for up to 1 week.

Decorative dough is prepared by combining the syrup with white rye flour and buckwheat flour. The ingredients are mixed until homogenous, shaped, and baked in a 300°F/149°C convection oven until dry. See the following chart for recipes to make plain and colored dough. The ratios are listed as bakers' percentages; see page 78 for more information on creating formulas using bakers' percentages.

DOUGH TYPE	WHITE RYE FLOUR	BUCKWHEAT FLOUR	SYRUP	COLOR
Plain Decorative Dough	75.1%	24.9%	63.8%	—
Black Decorative Dough	49.8%	40.3%	67.9%	10.0% cocoa powder
Red Decorative Dough	49.8%	40.3%	67.9%	10.0% chili powder
Brown Decorative Dough	75.0%	25.0%	56.8%	12.2% coffee extract
Yellow Decorative Dough	53.7%	43.4%	73.2%	2.9% turmeric

GLOSSARY

A

Acid A substance having a sour or sharp flavor. Foods generally referred to as acidic include citrus juice, vinegar, and wine. A substance's degree of acidity is measured on the pH scale; acids have a pH of less than 7.

Active Dry Yeast A dehydrated form of yeast that needs to be hydrated in warm water (105°F/41°C) before using it. It contains about one-tenth the moisture of compressed yeast.

Aerobic Bacteria Bacteria that require the presence of oxygen to function.

Agitate To stir.

Albumen A water-soluble protein found in egg whites.

Alkali A substance that tests at higher than 7 on the pH scale. Baking soda is an example of an alkaline ingredient.

Almond Paste A firm purée of ground almonds and sugar.

Anaerobic Bacteria Bacteria that do not require oxygen to function.

Ash Content The mineral content in flour.

Autolyse A resting period for dough after mixing the flour and water. This rest allows the dough to fully hydrate and to relax the gluten.

B

Bacteria Microscopic organisms. Some have beneficial properties; others can cause food-borne illnesses when contaminated foods are ingested.

Bakers' Percentage Method of calculating ingredient quantities, based on the total flour weight of a formula.

Baking Powder A chemical leavening agent composed of sodium bicarbonate, an acid, and a moisture absorber such as cornstarch. When moistened and/or exposed to heat, it releases carbon dioxide to raise a batter or dough.

Baking Soda A chemical leavening agent. Sodium bicarbonate is an alkali that when combined with an acid breaks down and releases carbon dioxide. This reaction causes the product to leaven as it is baked.

Banneton Basket made of plastic, straw, or willow that helps preserve the shape of a bread loaf during proofing.

Bâtard An oblong loaf of bread.

Bench Rest In yeast dough production, the stage that allows the preshaped dough to rest before its final shaping. Also known as secondary fermentation.

Biga Italian name for an aged dough. A type of pre-ferment containing 50 to 60 percent water and 0.3 to 0.5 percent instant yeast.

Blend To fold or mix ingredients together.

Boulanger The French word for "baker."

Boule A rounded loaf of bread.

Bran The tough outer layer of a grain kernel and the part highest in fiber.

C

Chemical Leavener An ingredient (such as baking soda or baking powder) whose chemical action is used to produce carbon dioxide gas to leaven baked goods.

Compressed Fresh Yeast This type of yeast is moist and must be refrigerated because it is extremely perishable.

Couche A heavy linen cloth used to protect loaves during proofing.

Crumb A term used to describe the interior texture of baked goods.

D

Decorative Dough Unleavened mixture of rye flour, buckwheat flour, and simple syrup used as a decoration on a baked good. Also known as dead dough or *pâte morte*.

Desired Dough Temperature The temperature of a finished dough that will produce the ideal rate of fermentation.

Dextrose A simple sugar made by the action of acids or enzymes on starch. Also known as corn sugar.

Disaccharide A complex or double sugar. When fructose and dextrose are bonded together, this is called sucrose, or table sugar. Maltose is another example of a disaccharide.

Dock To pierce dough lightly with a fork or dough docker (resembles a spiked paint roller) to allow steam to escape during baking. This helps the dough to remain flat and even.

Dough A mixture of ingredients high in stabilizers and often stiff enough to cut into shapes.

E

Emulsion The suspension of two ingredients that do not usually mix. Butter is an emulsion of water in fat.

Endosperm The inside portion of a grain, usually the largest portion, composed primarily of starch and protein.

Enriched Dough Dough that is enriched with ingredients that add fat or vitamins. Examples of these ingredients are sugar, eggs, milk, and fats.

F

Facultative Bacteria Bacteria that can survive both with and without oxygen.

Fermentation A process that happens in any dough containing yeast. It begins as soon as the ingredients are mixed and continues until the dough reaches an internal temperature of 138°F/59°C during baking. As the yeast eats the sugars present in the dough, carbon dioxide is released, which causes the dough to expand. Fermentation alters the flavor and appearance of the final product.

Folding The process of bending a dough over itself during the bulk fermentation stage to redistribute the available food supply for the yeast, to equalize the temperature of the dough, to expel gases, and to further develop the gluten in the dough.

Fondant Sugar cooked with corn syrup, which is induced to crystallize by constant agitation, in order to produce the finest possible crystalline structure. Fondant is used as centers in chocolate production or as a glaze in pastries.

Formula A recipe in which measurements for each ingredient are given as percentages of the weight of the main ingredient.

Fructose A monosaccharide that occurs naturally in fruits and honey. Also known as fruit sugar or levulose.

G

Gelatinization The process in which starch granules, suspended in liquid, are heated; they begin to absorb liquid and swell in size.

Germ The embryo of a cereal grain that is usually separated from the endosperm during milling because it contains oils that accelerate the spoilage of flours and meals.

Gliadin A protein found in wheat flour. The part of gluten that gives it extensibility and viscosity.

Glucose (1) A monosaccharide that occurs naturally in fruits, some vegetables, and honey. Also known as dextrose. (2) A food additive used in confections.

Gluten The protein component in wheat flour that builds structure and strength in baked goods. It is developed when the proteins glutenin and gliadin are moistened and agitated (kneaded). It provides the characteristic elasticity and extensibility of dough.

Glutenin A protein found in wheat flour. It is the part of gluten that gives it strength and elasticity.

Grain A seed or fruit of a cereal grass.

H

Homogenize To mix ingredients so they become the same in structure.

Hydrate To combine ingredients with water.

Hygroscopic Quality of absorbing moisture from the air. Sugar and salt are both hygroscopic ingredients.

I

Infuse To flavor by allowing an aromatic to steep in the substance to be flavored. Infusions may be made either hot or cold.

K

Kuchen The German word for "cake" or "pastry."

L

Lactose The simple sugar found in milk.

Lame Thin, arced razor blade clamped into a small handle, used to score proofed breads and rolls before they are baked.

Lamination The technique of layering fat and dough through a process of rolling and folding to create alternating layers.

Lean Dough A yeast dough that does not contain fats or sugar.

Leavening Raising or lightening by air, steam, or gas (carbon dioxide). In baking, leavening occurs with yeast (organic), baking powder or baking soda (chemical), and steam (physical/mechanical).

Liquefier An ingredient that helps to loosen or liquefy a dough or batter. Sugar, fats, and water or milk are examples of liquefiers in baking.

M

Maillard Reaction A complex browning reaction that results in the particular flavor and color of foods that do not contain much sugar, such as bread. The reaction, which involves carbohydrates and amino acids, is named after the French scientist who discovered it.

Mise En Place French for "put in place." The preparation and assembly of ingredients, pans, utensils, and plates or serving pieces needed for a particular dish or service period.

Mixing The blending of ingredients.

Monosaccharide A single or simple sugar and the basic building block of sugars and starches. Fructose, glucose, levulose, and dextrose are examples of monosaccharides.

O

Organic Leavener A living organism that ferments sugar to produce carbon dioxide gas, causing the batter or dough to rise. Yeast is an organic leavener.

Oven Spring The rapid initial rise of yeast doughs when placed in a hot oven. Heat accelerates the growth of the yeast, which produces more carbon dioxide gas and also causes this gas to expand. This continues until the dough reaches a temperature of 140°F/60°C.

P

Pain The French word for "bread."

Peel Wooden boards used to transfer dough, both to the loader and to and from the oven.

pH Scale A scale with values from 0 to 14 representing degrees of acidity. A measurement of 7 is neutral, 0 is most acidic, and 14 is most alkaline. Chemically, pH measures the concentration and activity of the element hydrogen.

Physical Leavening Occurs when air and/or moisture that is trapped during the mixing process expands as it is heated. This can occur through foaming, creaming, or lamination. Also known as mechanical leavening.

Polysaccharide A complex carbohydrate such as a starch, which consists of long chains of saccharides, amylose, and amylopectin.

Poolish A semi-liquid starter dough with equal parts, by weight, of flour and water that are blended with yeast and allowed to ferment for 3 to 15 hours.

Preferment A piece of dough that is saved from the previous day's production to be used in the following day's dough.

Preshaping The gentle, first shaping of dough. Also known as rounding.

Proof To allow yeast dough to rise.

R

Ratio A general formula of ingredients that can be varied.

Recipe A specific formula of ingredients and amounts.

S

Saccharide A sugar molecule.

Sanitize To kill pathogenic organisms chemically and/or by moist heat.

Scale Instrument to measure ingredients by weight.

Scaling Portioning batter or dough according to weight or size.

Score To make incisions into dough to allow steam to escape and the crust to expand. Also known as slashing or docking.

Secondary Fermentation See BENCH REST.

Soaker Grains and seeds that are soaked with water before mixing a dough.

Sponge Type of preferment commonly used for the production of sweet dough.

Stabilizer An ingredient that helps to develop the solid structure or "framework" of a finished product. Flour and eggs act as stabilizers in baking.

Starter A mixture of flour, liquid, and commercial or wild yeast that is allowed to ferment. The starter must be "fed" with flour and water to keep it active.

Steep To allow to infuse.

Sucrose Table sugar. A disaccharide extracted from sugarcane or sugar beets and consisting of glucose and fructose joined together in the molecule.

T

Texture The interior grain or structure of a baked product as shown by a cut surface; the feeling of a substance under the fingers.

BIBLIOGRAPHY

Alford, Jeffrey, and Naomi Duguid. *Flatbreads and Flavors: A Baker's Atlas.* New York: HarperCollins, 2008.

Anderson, Burton. *Treasures of the Italian Table: Italy's Celebrated Foods and the Artisans Who Make Them.* New York: William Morrow, 1994.

Assire, Jerome. *The Book of Bread.* Paris, France: Flammarion, 1996.

Bailey, Adrian. *The Blessings of Bread.* New York: Paddington Press, 1977.

Bonnefons, Nicolas de. *Les Delices de la Campagne.* Paris, France: L'Historia Plantarum, 1662. Reprint by Nabu Press, 2013, Paris, France.

Calvel, Raymond. *Le Pain et le Panification.* Paris, France: PUF, 1979.

Camporesi, Piero. *Bread of Dreams.* Chicago: University of Chicago Press, 1996.

Culinaria. *Culinaria: European Specialties.* New York: Könemann, 2005.

Culinary Institute of America. *Baking and Pastry,* 3rd Edition. New York: Wiley, 2009.

David, Elizabeth. *English Bread and Yeast Cookery.* London: Grub Street, 2011.

Dutton, Margit Stoll. *The German Pastry Bakebook.* Radnor, Pennsylvania: Chilton, 1977.

Field, Carol. *Celebrating Italy: The Tastes and Traditions of Italy as Revealed Through Its Feasts, Festivals, and Sumptuous Foods.* New York: HarperCollins, 1997.

Frazer, Sir James George. *The Golden Bough: A Study in Magic and Religion.* Oxford: Macmillan, 1900.

Galli, Franco. *The Il Fornaio Baking Book: Sweet and Savory Recipes from the Italian Kitchen.* San Francisco: Chronicle Books, 2001.

Giedion, Siegfried. *Mechanization Takes Command.* Milwaukee: University of Minnesota Press, 2014

Hamelman, Jeffrey. *Bread.* New York: Wiley, 2012.

Iaia, Sarah Kelly. *Festive Baking: Holiday Classics in the Swiss German and Austrian Tradition.* New York: Kirkus Service, 1988.

Jacob, H. E. *Six Thousand Years of Bread.* New York: Skyhorse Publishing, 2007.

Jaime, Tom. *Building a Wood Fired Oven for Breads and Pizzas.* England: Prospect Books, 2011.

Kaplan, Steven Laurence. *The Bakers of Paris and the Bread Question 1700–1775.* Durham, NC: Duke University Press, 1996.

Koehn, Walter A. *German Rye Bread Production with Rye Sour Dough.* White River Junction, VT: Chelsea Green Publishing.

Leader, Daniel, and Judith Blahnik. *Bread Alone: Bold Fresh Loaves from Your Own Hands.* New York: HarperCollins, 1993.

Montanari, Massimo. *The Culture of Food: The Making of Europe.* New York: Wiley, 1996.

Mennell, Stephen. *All Manners of Food: Eating and Taste in England and France from the Middle Ages to the Present.* Urbana-Champaign: University of Illinois Press, 1996.

Ortiz, Joe. *The Village Baker: Classical Regional Breads from Europe and America.* New York: Ten Speed Press, 2009.

Peterson, T. Sarah. *Acquired Taste: The French Origins of Modern Cooking.* Ithaca, NY: Cornell University Press, 1994

Poilane, Lionel. *Guide de L'Amateur de Pain.* Paris, France: Robert Laffont, 1994.

Romer, Elizabeth. *Italian Pizza and Hearth Breads.* New York: Clarkson N. Potter, 1987.

Root, Waverley. *The Food of Italy.* New York: Knopf Doubleday, 1992.

Sagarra, Eda. *A Social History of Germany 1648–1914.* Piscataway, New Jersey: Transaction Publishers.

Schunemann, Clus, and Gunter Treu. *Baking: The Art and Science.* New York: Baker Tech, 1988.

Toussaint-Samat, Maguelonne. *A History of Food.* New York: Wiley, 2009.

Wheaton, Barbara Ketcham. *Savoring the Past: The French Kitchen and Table from 1300 to 1789.* New York: Simon and Schuster, 2011

Ziehr, Wilhelm. *Das Brot.* Amsterdam: Atlantis, 1984.

APPENDIX: Mixing Log

MIXING LOG

DOUGH NAME	MIXING METHOD	ROOM TEMPERATURE	FLOUR TEMPERATURE	PREFERMENT TEMPERATURE	WATER TEMPERATURE	FINAL DOUGH TEMPERATURE

Conversion Charts

TEMPERATURE CONVERSIONS

Formula to convert from °F to °C: $\frac{{}^\circ F - 32}{1.8} = {}^\circ C$ (round up for .50 and above)

30°F = –1°C
32°F = 0°C
35°F = 2°C
38°F = 3°C
40°F = 4°C
45°F = 7°C
50°F = 10°C
55°F = 13°C
58°F = 14°C
60°F = 16°C
68°F = 20°C
72°F = 22°C
74°F = 23°C
75°F = 24°C [room temp]
77°F = 25°C
78°F = 26°C
79°F = 26°C
80°F = 27°C
82°F = 28°C
85°F= 29°C
86°F = 30°C
90°F = 32°C
95°F = 35°C
98°F = 37°C
100°F = 38°C
105°F = 41°C
110°F = 43°C
115°F = 46°C
120°F = 49°C
125°F = 52°C

130°F = 54°C
135°F = 57°C
140°F = 60°C
145°F = 63°C
150°F = 66°C
155°F = 68°C
160°F = 71°C
161°F = 72°C
165°F = 74°C
170°F = 77°C
175°F = 79°C
180°F = 82°C
185°F = 85°C
190°F = 88°C
195°F = 91°C
200°F = 93°C
205°F = 96°C
210°F = 99°C
212°F = 100°C
215°F = 102°C
220°F = 104°C
225°F = 107°C
230°F = 110°C
235°F = 113°C
240°F = 116°C
245°F = 118°C
250°F = 121°C
255°F = 124°C
257°F = 125°C
260°F = 127°C

265°F = 129°C
266°F = 130°C
270°F = 132°C
275°F = 135°C
280°F = 138°C
285°F = 141°C
290°F = 144°C
293°F = 145°C
295°F = 146°C
300°F = 149°C
305°F = 152°C
310°F = 154°C
315°F = 157°C
320°F = 160°C
325°F = 163°C
330°F = 166°C
335°F = 168°C
340°F = 171°C
345°F = 174°C
350°F = 177°C
355°F = 179°C
360°F = 182°C
365°F = 185°C
370°F = 188°C
375°F = 191°C
380°F = 193°C
385°F = 196°C
390°F = 199°C
395°F = 202°C
400°F = 204°C

405°F = 207°C
410°F = 210°C
415°F = 213°C
420°F = 216°C
425°F = 218°C
430°F = 221°C
435°F = 224°C
440°F = 227°C
445°F = 229°C
450°F = 232°C
455°F = 235°C
460°F = 238°C
465°F = 241°C
470°F = 243°C
475°F = 246°C
480°F = 249°C
485°F = 252°C
490°F = 254°C
495°F = 257°C
500°F = 260°C
505°F = 263°C
510°F = 266°C
515°F = 268°C
520°F = 271°C
525°F = 274°C
530°F = 277°C
535°F = 279°C
540°F = 282°C
545°F = 285°C
550°F = 288°C

WEIGHT CONVERSIONS

Formula to convert ounces to grams:

number of oz × 28.35 = number of grams (round up for .50 and above)

1 oz = 28.35 g = 28 g

Formula to convert pounds to kilograms

number of lb × 16 (10 × 16 oz = 160 oz)

multiply result by 28.35 and convert grams to kilograms:

160 × 28.35 = 4536 ÷ 1000 = 4.536 = 4.54 kg

½ oz = 14 g
⅔ oz = 19 g
¾ oz = 21 g
1 oz = 28 g
1¼ oz = 35 g
1⅓ oz = 37 g
1½ oz = 43 g
1⅔ oz = 47 g
1¾ oz = 50 g
2 oz = 57 g
2¼ oz = 64 g
2⅓ oz = 66 g
2½ oz = 71 g
2⅔ oz = 76 g
3¾ oz = 78 g
3 oz = 85 g
3¼ oz = 92 g
3½ oz = 99 g
3¾ oz = 106 g
4 oz = 113 g

4½ oz = 128 g
5 oz = 142 g
5½ oz = 156 g
6 oz = 170 g
6½ oz = 184 g
7 oz = 198 g
7½ oz = 205 g
8 oz = 227 g
8½ oz = 241 g
9 oz = 255 g
9½ oz = 269 g
10 oz = 284 g
10½ oz = 298 g
11 oz = 312 g
12 oz = 340 g
12½ oz = 354 g
13 oz = 369 g
13½ oz = 383 g
14 oz = 397 g
14½ oz = 411 g

15 oz = 425 g
15½ oz = 439 g
1 lb [16 oz] = 454 g
1 lb 4 oz = 567 g
1 lb 8 oz = 680 g
1 lb 12 oz = 794 g
2 lb = 907 g
2 lb 4 oz = 1.25 kg
2 lb 8 oz = 1.13 kg
2 lb 12 oz = 1.25 kg
3 lb = 1.36 kg
3 lb 8 oz = 1.59 kg
3 lb 12 oz = 1.70 kg
4 lb = 1.81 kg
4 lb 8 oz = 2.04 kg
5 lb = 2.27 kg
5 lb 8 oz = 2.50 kg
6 lb = 2.72 kg
7 lb = 3.18 kg
8 lb = 3.63 kg

9 lb = 4.08 kg
10 lb = 4.54 kg
11 lb = 4.99 kg
12 lb = 5.44 kg
12 lb 8 oz = 5.67 kg
13 lb = 5.90 kg
14 lb = 6.35 kg
15 lb = 6.80 kg
16 lb = 7.26 kg
17 lb = 7.71 kg
18 lb = 8.16 kg
19 lb = 8.62 kg
20 lb = 9.07 kg
22 lb = 9.98 kg
25 lb = 11.34 kg
30 lb = 13.61 kg
35 lb = 15.88 kg
40 lb = 18.14 kg
45 lb = 20.41 kg
50 lb = 22.68 kg

INDEX

Pages in *italics* indicate illustrations

Y

Z